# Elephant Sense and Sensibility

# Elephant Sense and Sensibility
## Behavior and Cognition

Michael Garstang

Photographers: Wynand du Plessis and Claudia du Plessis

AMSTERDAM • BOSTON • HEIDELBERG • LONDON
NEW YORK • OXFORD • PARIS • SAN DIEGO
SAN FRANCISCO • SINGAPORE • SYDNEY • TOKYO
Academic Press is an imprint of Elsevier

Academic Press is an imprint of Elsevier
125 London Wall, London, EC2Y 5AS, UK
525 B Street, Suite 1800, San Diego, CA 92101-4495, USA
225 Wyman Street, Waltham, MA 02451, USA
The Boulevard, Langford Lane, Kidlington, Oxford OX5 1GB, UK

**Notices**
Knowledge and best practice in this field are constantly changing. As new research and experience
broaden our understanding, changes in research methods, professional practices, or medical
treatment may become necessary.

Practitioners and researchers must always rely on their own experience and knowledge in
evaluating and using any information, methods, compounds, or experiments described herein. In
using such information or methods they should be mindful of their own safety and the safety of
others, including parties for whom they have a professional responsibility.

To the fullest extent of the law, neither the Publisher nor the authors, contributors, or editors,
assume any liability for any injury and/or damage to persons or property as a matter of products
liability, negligence or otherwise, or from any use or operation of any methods, products,
instructions, or ideas contained in the material herein.

**British Library Cataloguing-in-Publication Data**
A catalogue record for this book is available from the British Library

**Library of Congress Cataloging-in-Publication Data**
A catalog record for this book is available from the Library of Congress

ISBN: 978-0-12-802217-7

For information on all Academic Press publications
visit our website at http://store.elsevier.com/

Typeset by SPi Global, India

Printed and bound in The United States
15   16   17   18        10   9   8   7   6   5   4   3   2   1

Working together
to grow libraries in
developing countries

www.elsevier.com • www.bookaid.org

# Contents

# Foreword

A million years ago, a diverse array of elephants and their probisidian kin occupied all of the continents except Australia and Antarctica over a wide array of different habitats. Today, they are found in sub-Saharan Africa and parts of Asia. Multiple theories postulate the cause of the relatively recent demise of this widespread, successful group of animals. Many archeologists and paleo-ecologists indict human hunting pressures as a root cause, but competing theories abound. Whatever the case may be for the past, there is little question that modern humanity now manifests a horrific toll on the remnant elephant populations. Tragically, much of the current pressure is propelled by elephant poaching for trinkets—bits of carved ivory as *objets d'art* and jewelry; entire tusks as symbols of different aspects of power and importance. Economics, the dismal science, postulates as the rarity of a commodity increases, so too does its price. This greatening incentive for poachers to illegally harvest diminishing elephant herds for increasingly valuable ivory augers poorly for the future.

As a remedy to this gloomy view, Michael Garstang—with the help of the spectacular photographs of his collaborators, Wynand and Claudia du Plessis—illuminates the wonder of the African elephant, *Loxodonta africana*, in multiple dimensions in this book. The amazing morphology of elephants includes their impressive size since they are the largest animals to now walk the earth. In the case of the African elephant, the trunk has the equivalent of opposable digits, the same adaptations that differentiated our species from its near kin and allowed complex manipulation of objects and eventually tools. Indeed, the elephant also has a well-developed, large brain with a considerable portion emphasizing the same parts of the human brain associated with higher thought processes. Along with the physical adaptations in the elephant, there is an equal array of sensory adaptations. The elephant's trunk combines dexterity with a keen sense of touch and smell in an organ that can both drink and expel water. The trunk of an elephant is involved with sound production, lifting hundreds of kilograms or grasping and inspecting a leaf. The unrelated tapirs or the extinct and really unrelated large marsupials in Australia, *Zygomaturus trilobus et al.*, feature prehensile trunks, but nothing compared to the trunk of an African elephant. The multiple uses of organs seem a standard feature of elephants, with their ears growing larger flaps to shed heat in hotter environs and also hearing long-wave sounds that most other animals cannot. As he discusses in his text, Professor Garstang was an early discoverer of the capacity of elephants to communicate over tens of kilometers, especially when meteorological conditions are right.

There is an anecdote on appreciating the importance of understanding things as a whole that tells of four blind men encountering and describing an elephant. One touches its trunk and says that an elephant is like a hose; another touches its side and reports the elephant resembles a wall; touch the tail and it is like a rope; touch the leg and it is like a tree. Directly true to parable, elephants are made of remarkable parts but they are much more *in toto*: as an organism; as a herd; as a population; as a controller of ecosystem pattern. Elephants are complex creatures. So much so that the young have to learn how to be elephants. The rich adaptations seen in their behaviors are a strong part of this book. The altruism of the herd, the education of young elephants, and the richness of their behavior would not be experienced by the four blind men of the parable nor by almost anyone else unless it is pointed out. Professor Garstang's text does that and more.

Michael Garstang brings a remarkable experiential kit to his discourse on the African elephant. His boyhood in northern Kwa-Zulu in Natal, Republic of South Africa, gave him a deep and personal appreciation of elephants and the ecosystems of South Africa written large. It also gave him Zulu as a second language. He later interwove elephant behavior and Zulu legend to produce a wonderful children's book of the story of a young elephant, Ntombazana, and her life in the same locale in which Garstang himself grew up. Another book, *Observations of Surface-to-Atmosphere Interactions in the Tropics*, reflects his scientific research. He is a scientist with a doctorate in meteorology and an enviable publication and professional reputation across his field. Dr. Garstang was appointed the 2012 Artist-in-Residence at the Virginia Coast Reserve Long-Term Ecological Research Station in Oyster. He is a man of many talents and dimensions, all of which coalesce in the present book *Elephant Sense and Sensibility*. T.S. Eliot noted, "We must know all of Shakespeare's work in order to understand any of it." Likewise, Dr. Garstang's diverse intellectual toolkit is the prerequisite of understanding enough of the African elephant to understand any of it.

**H.H. Shugart**
*W.W. Corcoran Professor of Natural History*
*Department of Environmental Sciences*
*University of Virginia*

## A SPECIAL NOTE ON THE PHOTOGRAPHY

All of the photographs, with the exception of one by the author of the Nyala antelope on page 84, are the work of Wynand and Claudia du Plessis (www.claudiawynandduplessis.com; www.wildphoto-shop.com; www.photo-art-prints.com). Born in South Africa and Germany, respectively, Wynand and Claudia are trained in ecology, environmental geography, and wildlife management. They spent 10 years in the Etosha National Park in Namibia and are now accomplished wildlife photographers, publishing a book titled *Etosha: Rhythms of an African Wilderness* and many images for international media including National Geographic, Oprah, and Mercedes. They have won multiple awards at the BBC/BG Wildlife Photographer of the Year competition.

# Preface

The Bells of Heaven

*'Twould ring the bells of heaven,*
*The wildest peal for years,*
*If Parson lost his senses*
*And people came to theirs.*
*And he and they together*
*Knelt down with angry prayers*
*For tamed and shabby tigers,*
*And dancing dogs and bears,*
*And wretched, blind pit ponies,*
*And little hunted hares.*

Ralph Hodgson, 1927.

To sense, before you see, the shape of an elephant emerging soundlessly from the gloom of an African night is to make tangible contact with the soul of Africa. Here before you is an animal weighing more than 7 tons and perhaps twice or three times your height. This alone causes you to gasp. Yet it is but the beginning of a story that is full of mystery and surprises.

It is the purpose of this book to examine in both words and images the core being of an elephant. To use what is known and what is only suspected to describe how elephants deal with the world they occupy, to document not how we think elephants think but how to think as elephants.

While we are shackled to our human perceptions, we make every effort in this book to see the world through the senses of an elephant. Verbal description of complex behavior provides the framework for visual representation of those responses. A lifetime of experience is called upon to capture visual images and to match these with behavioral states.

The accumulated evidence presented in this book shows elephants as sentient beings differing from humans not so much in kind but in degree. If elephants think, act, feel, and behave in ways we as humans recognize, how do we justify our dominion over these animals? The exercise of this privilege by humans over animals becomes ever more evident as human populations continue to expand and less and less space is conceded to animals. Species are lost, contact with the natural world is diminished, and life becomes an artifact of our imaginations. We are left with a cyber world to fill the void of lost species and lost spaces. The hope within this vision of a sterile world is that despite conflict between humans and animals, and in particular between humans and elephants, Darwinian survival and evolution will prevail. Elephants can share this planet with humans if only humans will allow them to do so.

To a large degree humans have separated themselves from the rest of the animal world. Philosophers as far back as Aristotle have debated whether animals possess emotions in the same sense that humans do. Darwin saw social interests as the bonds that hold a group together and represent the beginnings of morality. A number of present-day behavioral biologists see morality in evolutionary terms with altruism, reciprocity, trust, empathy, compassion, sharing, and a sense of fair play preceding by millions of years any human conception of these emotions.

A minority of present-day philosophers see animals as moral subjects who can act on the basis of moral reasons. It is argued that moral reasons stem from moral emotions, which are intentional states that possess identifiable moral content. The ultimate conclusion is that if animals can act for moral reasons then they are worthy of moral respect.

Most philosophers, however, from Aristotle to Kant to Korsgaard, think that we humans created "the order of moral values" and decided to "ratify and endorse the natural concern that all animals have for themselves" even though it is a "condition shared by other animals."

Is it not possible that it happened the other way around? Long before the appearance of any humanoid animals on this planet, animals existed who mattered to themselves, who pursued their own good, and who thus constituted value as subsequently and much later recognized by philosophers. It did not take the emergence of humans to establish these values. They existed whether codified or not.

Even though value is seen as a human creation, made both possible and necessary by human rationality, the basis for moral behavior rests upon behavior that promoted survival. Elements of such behavior are present in all species, most directly in the relationship between mother and child. Such behavior multiplies in species that demand protracted motherly care and is extended progressively in social species to reach our current perception of such behavior in a species capable of communication and rational thought. While admitting that language and rational thought have refined moral codes, failure to do so does not eliminate morality.

Sense in the English language is a complex word requiring in almost any dictionary at least half a column of space to define. In general, sense is the perception using sight, hearing, smell, taste, and touch of the world around us. Sense can be "a feeling of consciousness" generated by one or more of the senses, whereas sensibility is the capacity to translate that vagueness of feeling into an emotional feeling. In the case of elephants, as opposed to humans, it may be possible that these animals are in many instances more finely tuned to detect and translate more than one signal and that their sense and sensibility may equal or even exceed ours. It is this capacity that will form the core of this book.

**Michael Garstang**

# About the Author

Born and raised in remote northern KwaZulu-Natal, South Africa, Michael Garstang claims a distinguished career in meteorology, publishing groundbreaking findings on the role of the tropics in the global atmosphere, discovering how Saharan dust sustains the rain forests of Amazonia, and finding out how the atmosphere plays a crucial role in elephant survival. Dr. Garstang is a Distinguished Emeritus Research Professor in Environmental Sciences at the University of Virginia.

Elephant at waterhole at Etosha National Park, Namibia (water color by author).

# Acknowledgments

The wide-open spaces of the Mzinyathi Plains and the towering kranzes of the Drakensberg, together with Zulu folklore and legends, laid the foundation for my enduring fascination with the natural world. For these early beginnings I am eternally grateful and thank the known and unknown spirits who gave me these gifts.

For those who have looked down on Africa from a high-flying aircraft, the overwhelming sensation is that this is where the world began. Southern Hemisphere geophysicists and paleontologists such as du Toit, Dart, Broome, and Leaky brought the "bushman" paintings on the sandstone cliffs to life. Family history of elephant herds together with early books on elephants by Iain and Oria Douglas-Hamilton, Moss, Poole, and Payne consolidated my fascination with these great mammals.

Opportunities and knowledge granted me by Joanne and Robert Simpson and Noel LaSeur, all of tropical meteorology and hurricane fame, established my career and subsequent life.

Like any academic, classmates, colleagues, and especially students who taught me more than I taught them are an integral part of my life. Of the many, the few—Ross Houghton, Ed Zipser, Don Brown, David Fitzjarrald, and Bob Davis—must serve to represent the many. Together with Mary Morris, my assistant for many years, go my eternal but inadequate thanks.

My wife, Elsabé, made graduate school possible, raised a family and has over six decades supported whatever current all-consuming project I indulged in. Thanks are an inadequate expression of my gratitude.

Chapter 1

# Introduction

The Namib Desert, stretching inland along the west coast of southern Africa, contains the largest dune fields of all the world's deserts. Fixed dunes rising to over 100 m (320 ft) form formidable barriers between the interior of Namibia and the Skeleton Coast. Despite this, desert elephants (Figure 1.1), tallest of the African savanna elephants (*Loxodonta africana*), cross these dunes to reach isolated oases to drink and feed on preferred vegetation.

The Angolan civil war (1975–2002) decimated and dislocated wildlife in this region. It is estimated that 100,000 elephants were exterminated in Angola alone. Wildlife in Namibia was also disrupted by military operations and by poaching. A north–south road, built for military purposes in Namibia, cut off migrations from the east (Etosha National Park) to the coastal deserts of north-western Namibia.

With independence in Namibia and a fragile peace in Angola (Figure 1.2), the blocking road was closed. Elephants could once again penetrate the desert. Reaching the nearest oasis meant a 24-h walk, across shifting desert sands, climbing up and over 100-m (300-ft) dunes in a featureless terrain devoid of landmarks. Elephants who had not made this journey for more than two generations unerringly crossed these sands to revisit favored isolated sites. Whether or not the elephants that made this amazing journey had done so before still leaves the questions: How did they navigate across such hostile and featureless terrain to find a precise and isolated location? And, how and why did they remember these remote clusters of green in a vast sea of undulating sand?

Similar journeys have been recorded covering some 180 km (112 mile) from the Caprivi Strip and Botswana by elephants returning to Angola's Luiana Partial Reserve (Leon Marshall, January 2008, *Sunday Independent* [South Africa]). This reserve in Cuando Cubango Province was occupied by Jonas Savimbi's National Union for the Total Independence of Angola (UNITA). Savimbi's rebels distributed unknown numbers of landmines in the region. Michael Chase of Elephants Without Borders (Chase, 2007; Chase and Griffin, 2011) knows of the location of some 45 minefields near Jamba Camp, Savimbi's old headquarters. After the end of the civil war in 2002 when elephants began returning to the Luiana Partial Reserve, many were fatally injured by these land mines. Chase, however, reports that within 2 years no incidents of injury or loss of elephants in the region were reported. By using overlapping tracking of five

Elephant Sense and Sensibility. http://dx.doi.org/10.1016/B978-0-12-802217-7.00001-6

**FIGURE 1.1**    Desert elephant, *Loxodonta africana*, adapted to the sands of the Namib and Kalahari. Taller and more slender than its fellow elephants on the wetter savannas, the tallest on record is 4.5 m (14 ft, 4 in.) at the shoulder.

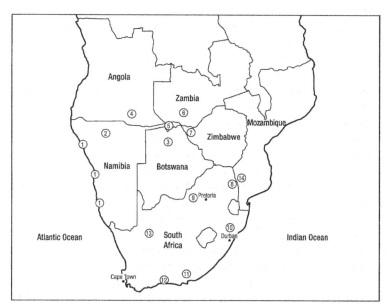

**FIGURE 1.2**    Points of interest in Southern Africa that appear throughout the text: (1) Namib Desert and Skeleton Coast; (2) Etosha National Park; (3) Okavango Delta; (4) Luiana Partial Reserve; (5) Caprivi Strip; (6) Luangwa National Park; (7) Sengwa National Park (formerly Wanki); (8) Kruger National Park; (9) Pilanesberg National Park; (10) Hluhlwe Game Reserve; (11) Addo Elephant National Park; (12) Knysna Forest; (13) Kalahari; and (14) Parque National do Limpopo.

GPS-collared elephants, Chase demonstrated that these elephants (both collared and herd mates) were moving through minefields without injury. It would be difficult not to conclude that the elephants had learned to detect the presence of mines, knew that the mines represented danger, and were able to avoid them. It is equally likely that these elephants used their highly refined sense of smell to detect the mines.

Responding to the changing circumstances described above, elephants demonstrate cognitive abilities and adaptability that are remarkable. The ability to find remote locations in trackless landscapes, to deal with threats to their survival, and to formulate solutions that can be followed by the group as a whole draw upon advanced mental processes. Elephants depend on memory, making the origins of memory fundamental to elephant neural processes. Memory is thus a central theme of this book. We as humans are what memory makes us. Without memory we cease to exist as sentient beings. This is no less true for elephants than it is for humans.

In Chapter 2 we trace the evolution of elephants with particular attention to aspects of the elephant's brain that reflect this evolution. Elephants deviated from the primates some 35 million years ago. Yet elephants, proceeding in parallel, evolved brains that are functionally more similar to those of humans than they are different. With large bodies and complex systems such as the trunk and an opposable thumb in the form of a highly tactile and sensitive tip (Figure 1.3), elephants exhibit cognitive abilities that may in some instances exceed those of humans. Because human brains have been studied much more extensively than those of elephants, in Chapter 4 we make some comparisons between these two brains.

In Chapter 3 we examine physiological aspects of the elephant's brain but in concert with current neurological research (Bryne and Bates, 2006); we are more concerned with how the brain as a whole functions rather than the role played by the component parts of the brain. In particular, we attempt the difficult task of entering the mind of the elephant, despite the constraint of the imbalance of our knowledge of the elephant versus the human brain (Gould, 2004).

We take a Darwinian approach, arguing that evolution favors behavior that promotes survival of the species (Darwin, 1897). This will lead us to explore in elephants the existence and the origin of memory (Chapter 5), morality (Chapter 6), emotions (Chapter 7), empathy and altruism (Chapter 8), communication (Chapter 9), language (Chapter 10), intelligence (Chapter 11), learning and teaching (Chapter 12), the sensory environment (Chapter 13), and the relationship between humans and elephants (Chapter 14). Each of these characteristics is embedded in the cognitive processes of the elephant's brain and is thus uniquely elephantine.

We draw upon the growing body of scientific evidence that examines these areas and while we are able to consider significant findings, these will not be without contention, nor in many cases fully supported by definitive scientific

**FIGURE 1.3**    The versatility of the trunk is enhanced by the two fingers or the equivalent of an opposable thumb at the tip of the African elephant's trunk. The sensitivity and control of this prehensile feature allows the detection of minute surface features or the picking-up of objects as small as an unshelled peanut.

evidence. Few animals can be subjected to rigorous and controlled experiments. This is not only for ethical reasons but as we show, failure to replicate natural conditions can invalidate or bias many experiments that are carried out under artificial conditions. Conversely, however, it is difficult to conduct rigorous experiments under presumed "natural" conditions. In many instances, we draw upon unverified anecdotal evidence not as proof of a concept but as possible guidance to the development of a testable hypothesis or the formulation of a question (Byrne, 1997; Byrne and Bates, 2011a; Heyes, 1993).

In conclusion, we ask whether our findings cast light upon the relationship between ourselves and our fellow beings who occupy this planet with us, addressing the troubling question of what endows humans with significant privileges over animals.

# Chapter 2

# Elephant Evolution

Darwinian evolution, alluded to in the previous chapter, is seen in contemporary biology in complex and sometimes conflicting terms. We subscribe by and large to the views on evolution as developed by Richard Dawkins in both *The Selfish Gene* (1989) and *The Extended Phenotype* (1999). Our approach, although predicated upon the following discussion of how Dawkins might view evolution, will in many cases point only to a simplified conception of how crucial characteristics of elephant behavior might have evolved, without entering into the complexities of how or perhaps even whether, in fact, this may have taken place.

Dawkins (1989) holds that "the fundamental unit of selection, and therefore of self interest, is not the species, nor the group, nor even, strictly, the individual. It is the gene, the unit of heredity" (p. 11). Dawkins sees natural selection leading to stable forms of life with high longevity, fecundity, and copying fidelity coded within the DNA of the individual (see pp. 18–23, 41). Given that life begins with the gene, aggregating into what are referred to as phenotypes and residing in a transfer mechanism he describes as a replicator, Dawkins (2008) does not neglect nor deny the existence of complex individual organisms. He sees the organism as a physically discrete machine, with essential internal organization. It is a definable unit, with a unique collection of the same genes. Yet, it is different from all other organisms. It has its own coordinated central nervous system such that "all its limbs conspire harmoniously together to achieve one end at a time" resulting in "intricate orchestration, with high spatial and temporal precision, of the hundreds of muscles in the individual" (pp. 250–251).

Copying errors will occur, resources will be stressed, and competition can delete the less favored, but nothing controls the process. "Trivial tiny influences on survival probability can have a major impact on evolution. This is because of the enormous time available for such influences to make themselves felt" (Dawkins, 1989, p. 4).

Evolution does not act for the good of the species but for the survival of the individual (perhaps "selfish") gene. Yet these self-preserving, genetically inherited traits under certain circumstances might serve to promote that society. For this to happen, Dawkins claims "that we shall be faced with something puzzling, something that needs explaining" (p. 4). The male may be banished from the group and be forced to compete and distribute his selfish genes among many different groups. These groups, guided by a single female who occupies

Elephant Sense and Sensibility. http://dx.doi.org/10.1016/B978-0-12-802217-7.00002-8

her position not through conflict but by an approximation to acclimation, not only nurtures her young for a long period of time but retains a high proportion of her female offspring within a single group for their entire extended lifetime. Perhaps this is the unusual circumstance that is needed for the selfish gene to contribute or be displaced in order that the species survives?

A significant part of the content of the pages that follow documents how elephants might be the unique species of animal that most closely meets the unusual circumstances demanded by an evolutionary mechanism centered upon the propagation of genes.

Proboscideans, the ancestors of today's elephants, can be traced back in time for more than 35 million years (Figure 2.1). They evolved in dense, closed-canopy forests. By the time that Elephantidae in the form of the mammoth and present-day Asian (*Elephas maximus*), African savanna (*Loxodonta africana*), and forest (*Loxodonta cyclotis*) elephants emerged as distinct species, the forests had receded, giving way to savannas. Because of this evolution within a forest, elephants emerged onto the open savannas with capabilities formed in the forests. Their sense of hearing and smell had evolved in favor of sight. Elephants were thus capable of both emitting and detecting calls over the widest range of frequencies of any animal. Their eyesight was poor, adjusted more to the restrictions of dense vegetation than to the vistas of the savannas.

Dawkins (1989, p. 7) relates a story told by Colin Turnbull who took a pygmy friend, Kenge, out of the forest for the first time in his life. They climbed a mountain and had an extended view over the plains where in the far distance a herd of buffalo were grazing. Kenge turned to Turnbull and asked, "What insects are those?" Puzzled by the question, it took Turnbull a moment to realize that in the limited vision of the forest there was no need to adjust for distance when judging size. Without any experience of using known objects as a basis for comparison, Kenge was unable to interpret the size of what he saw.

The vertical structure of temperature, moisture, and wind in the forest favored long-range transmission of low-frequency sounds with wavelengths on the order of meters to tens of meters rather than centimeters. Trees and vegetation have little attenuating affect on sounds with such long wavelengths. Wind speed and turbulence, which effectively destroy sound and limit the distance over which sound travels, are low to nonexistent in the forest. Temperatures at the cool floor of the forest are, especially during the day, much lower than at the sunlit forest canopy, resulting in an increase in temperature from the floor to the tops of the trees. This phenomenon, called a temperature inversion, results in denser air (lower temperatures) at the surface and less dense air (higher temperatures) at the tops of the trees. Sound waves produced at the surface in such a layer of air are bent first upward and then downward as their speed changes with lower and higher temperatures. The sound wave effectively bounces down this inversion channel. This ducting of sound allows the calls of elephants to be heard by other elephants at distances of kilometers. A similar temperature gradient is present in the world's oceans, where it is known as the SOFAR channel

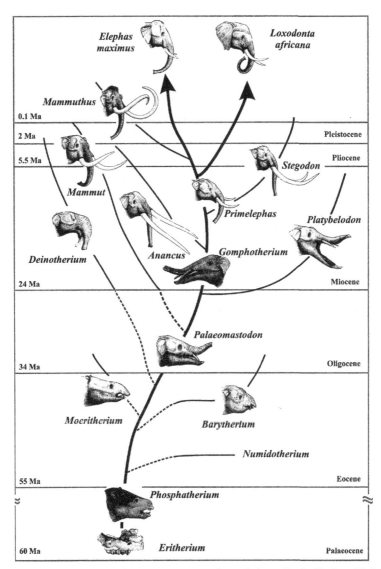

**FIGURE 2.1** Simplified diagram of the emergence of true elephants (family Elephantoidea) (after Shoshani, 2002). Red lines show the geologic periods and associated times. The heavy continuous black line shows the evolution of *Elephas maximus and Loxodonta africana* (*Loxodonta cyclotes* not shown). The subsidiary lighter/dashed lines show some but not all of the extinction of families. *Credit and permission derived from Vladimir Nikolov and Dr. Docho Dochev, Department of Geology, Paleontology and Fossil Fuels, Sofia University, Sofia, Bulgaria.*

or thermocline and is used by marine mammals such as whales to send signals to conspecifics thousands of kilometers away. The utility of this ability to communicate over vast distances can be measured not only in terms of finding mates but of being able to use food resources separated by many hundreds of miles of barren ocean. The ability not only to know where food is but to have unobstructed access to these locations is something that terrestrial animals are being progressively denied as their habitats shrink and migration routes are severed.

In a rather surprising twist of fate, the dry savannas that most of the elephants now found themselves occupying exhibit a powerful night-to-day reversal in acoustic conditions. Cloud-free, dry atmospheres of the savannas allow the daytime heat gained through absorption of solar radiation to stream upward and outward to space as soon as the solar angle drops toward sunset. The loss of outgoing longwave or terrestrial radiation operates to rapidly cool the earth's surface. Surface temperatures that may have approached 45–50 °C (113–122 °F) in the middle of the day and early afternoon now drop precipitously to 10 °C (50 °F) or lower, producing a daily range in temperature of some 40 °C (72 °F), more than many locations on earth experience from the hottest summer day to the coldest winter night. A very strong and shallow nocturnal inversion is formed, allowing loud elephant calls to be heard by another elephant as much as 10 km (6 mile) away. This ability to communicate over such great distances translates to being heard over an area of greater than 300 km$^2$ (112 mile$^2$). Such a reach in communication is crucial to elephant reproduction, predator avoidance, and resource utilization, all factors that we examine in terms of their behavior, social structure, and cognition (Garstang et al., 1995, 2005; Larom et al., 1997).

Conversely, during the day when surface temperatures rise above 40 °C, plumes of rising air generate turbulence and turbulent mixing. Because winds above the surface are always stronger than at the surface, these turbulent eddies bring higher winds to the surface, exacerbating the turbulence and attenuation of sound. Under the worst daytime conditions, elephant calls that traveled 10 km (6 mile) and covered 300 km$^2$ (112 mile$^2$) during the early evening can now only be heard 1 km (0.6 mile) away and only over an area as small as 3 km$^2$ (1.3 mile$^2$). Finding a mate, avoiding a predator, or effectively using sound to share food resources are all seriously impaired. Research shows that elephants on these savannas, in common with a number of other animals, rarely use their loud, low-frequency calls during the middle of the day (Garstang et al., 2005).

Predisposed by their evolution in forests, elephants are equipped with unusually effective sound-producing and -detecting systems coupled with extraordinary olfactory sensors. Sound and smell for these species play a fundamental and pervasive role in their day-to-day lives and in the ultimate evolution as a species. When asking how an elephant's mind works, these evolutionary underpinnings must govern much of our thinking about elephant cognitive abilities.

In our attempt to understand the mind of an elephant, we must think in terms other than those that guide the primates. Not only will this reveal evolutionary parallels between two quite divergent species but will surely provide us with a better understanding of the cognitive powers of this great animal.

# Chapter 3

# An Elephant's Brain

Current neural research has tended to move away from seeking to assign specific functions to anatomically different parts of the brain. Evidence of the interconnectivity of the brain, the degree of feedback between different parts of the brain, and the speed and number of signals transmitted over very short time periods (seconds and less) all suggest that only limited understanding can be gained from the examination in isolation of any single part of the brain. Intelligence, which has always been difficult to define, now seems to reside everywhere and nowhere (Holdrege, 2001).

The complexity of the brain is illustrated by our struggle to make computers simulate the brain. Early optimism in the wake of the development of computers (1960s) promised artificial intelligence that would begin to challenge humans. The multiple feedback loops operating in the brain could not be simulated by simple input–output devices. It was not until 50 years later when computers could be programmed to make choices based on statistical probabilities that an approach to "thinking" was achieved and demonstrated by the performance of IBM's Watson on the television program *Jeopardy* in 2011.

Having taken due note of the difficulties inherent in trying to think like an elephant and avoiding reliance upon a reductionist approach, we can, with caution, draw some benefits through the comparison of component parts of the human and elephant brains.

The elephant's brain is three to four times larger than a human brain, weighing an average of 5.0 kg (11 lbs) compared to 1.45 kg (3.8 lbs) for a human brain. Elephant female brains are slightly smaller than those of males (4.677 kg or 10.3 lbs). This difference is displayed in the brains of humans (Shoshani et al., 2006). The heaviest known elephant brain on record weighed in at 9.0 kg or 19.8 lbs. When the brain is scaled to the body weight, the ratio is 1/600–1/700 for the elephant compared to 1/50 for a human and 1/40 for a dolphin. The encephalization quotient (EQ), which is the ratio of the observed brain size to the average brain size for the weight of the animal such that the average EQ = 1, has been used as a rough guide to intelligence. With EQ less than 1, intelligence is assumed to be less than average. With an EQ greater than 1, intelligence is assumed to be above average. Humans have EQs between 7.33 and 7.69, the next highest are chimpanzees, with EQs between 2.18 and 2.49. Elephants at the top end of their EQ scale (2.36) rank higher than the rest of the great apes, monkeys,

Elephant Sense and Sensibility. http://dx.doi.org/10.1016/B978-0-12-802217-7.00003-X

lemurs, and other mammals measured. Intelligence, as stated above, remains difficult to define. Shoshani and Eisenberg (Holdrege, 2001) define intelligence as the "capacity to meet new and unforeseen situations by rapid and effective adjustment of behavior." We progressively identify here behavior that can be related to intelligence.

The brain of an elephant calf at birth may be as high as 53% of its final size, but it may not reach full weight before the 15th year. Human brains at birth are 25% of their final weight, doubling in the first year and by the 6th year have reached 90% of final weight. The final weight of the human brain is reached in the 16th or 17th year. Brain size is certainly related to body size, motor skills, and the environment. The size of an elephant demands extensive motor controls. The tallest modern-day elephant, found as a skeleton in Namibia, measured over 4 m (14 ft) at the shoulder. These tall, long-legged desert elephants are not the heaviest. The heaviest bull elephant, the African savanna elephant, probably weighs over 7 ton (14,000 lbs). Yet these giants can move through thick bush or forest in deathly silence. Elephants tend to use favored game trails, which can be centuries old (early aviators in Africa found game trails a reliable means of navigation in trackless regions of the continent). The surface of these trails is often a mixture of sand and fiber from centuries of droppings, producing a soft, spongy surface that can be trodden in silence. Elephants exercise considerable control over vegetation. By breaking branches and pushing over trees, they serve as a keystone species keeping savannas open to grasslands and supporting ungulate populations (Figure 3.1). Elephants have been termed *mega-gardeners*. Campos-Arceiz and Blake (2011) estimate that African elephants disperse seeds from at least 335 plant species and 213 genera. The acute olfactory sensors of elephants are used to locate seeds and fruit. Excellent spatial and temporal memories enable them to locate fruit when ripe. Savanna elephants are estimated to distribute over 2000 seeds per square kilometer every day. Some 15 tree species have been found dispersed by elephants over distances as great as 50 km (31 mile). Such dispersal is far more successful than germination around the parent tree. Seeds dispersed by elephants germinate 57% of the time, whereas only 3% of the seeds fallen under the parent tree germinate.

In the Salonga National Park in the Democratic Republic of the Congo, Beaune and colleagues (2013) found that seeds from most plant species are dispersed by animals rather than by wind, water, or ballistic mechanisms. Elephants and bonobos were found to be the largest seed dispersers. They estimated that 85% of all plant species and 88% of tree species but 95% of individual trees were dispersed by animals in the forest that they studied. The loss of animal species and, in particular, elephants could result in a radical change in the composition of the forest.

Evidence of the exquisite and delicate motor control of these giant animals can be observed at almost any time in watching a family group. Mothers and aunts, weighing 5 ton and towering over their newborn calves who they are unable to see under their wide bodies, can be gently nudged onto their feet or

**FIGURE 3.1**   The strength of the trunk, with some 1400 muscles, is considerable. It can be used to break branches by twisting or bending, push over large trees, and, with the aid of its tusks, strip bark and gouge out softer fibrous materials.

moved aside by legs as large as tree trunks and feet the size of dining chair seats. At water holes, it is not unusual to see a full-grown elephant gently step over a tortoise no bigger than a small dinner plate.

At the Elephant Sanctuary in Hohenwald, Tennessee, an adult female elephant (Tarra) had made firm friends with a dog named Bella. Bella liked to have her tummy rubbed. Tarra would oblige using a front foot to do so. Tarra's front foot was half the size of Bella (Buckley, 2009).

The trunk of an elephant contains no bone or cartilage, giving this unique organ an almost unlimited range of motion all coordinated by muscle movements. The trunk is in almost continuous use while feeding (Figure 3.2). It is used to select parts of plants, clean food, fold and manipulate food into rolls or bundles, even use fibrous material to stop up holes dug for water, access water in deep fissures, spray water and mud over the body, snorkel, throw objects, and manufacture and manipulate tools. The prehensile tip can pick up objects as small as an unshelled peanut and detect cracks or grooves no more than a quarter of a millimeter wide. The elephant can use the strength of its trunk to push over sizable trees, strip bark off trees using both its trunk and its tusks, lift logs perhaps

**FIGURE 3.2**   The trunk is in almost continuous use as an effective organ to gather the amount of vegetation to meet the daily needs of an adult elephant.

as heavy as half a ton, and use the trunk as a deadly weapon that can disable a full-grown lion or kill a human. Finally, the trunk of an elephant as part of its long vocal tract may be critical to the production of loud, low-frequency calls as well as serving as an organ extremely sensitive to touch and to smell. Mental control of this complex organ occupies a large fraction of the elephant's brain (Shoshani et al., 2006).

The cerebellum of the elephant's brain is proportionally larger than a human's (18.6% of the brain vs. 10.3%) with gyri and sulci that have complicated folia and many subconvolutions that are all related to coordinating movements, locomotion, and posture. The *large cerebellum* and *basal ganglia* favor the ability to perform complex and coordinated motor functions like those carried out with the trunk.

In humans, the temporal–parietal junction gathers information from many senses (visual and tactile) to construct a single image. It is unknown but likely that elephants have a similar capability (de Waal, 2013, p. 92).

The large and bulbous *temporal lobes* including *anterior temporal gyrus*, together with a well-developed *neocortex* and large *olfactory bulb* and *paleopallum*, all support an excellent ability to smell. The olfactory region of an elephant's brain includes complex functions such as flehmen (reproduction) and Jacobson's (vomeronasal) organ housed in the roof of the mouth as well as a highly tuned sense of smell in the trunk. The very large *temporal lobe* of the cerebrum of the elephant may be related to the complex forms of communication including infrasound used by the elephant.

Elephants and humans have the highest degree of convolutions in the brain of the 13 mammalian taxa. The relative weight and surface area of the elephant's *cerebral cortex* and *temporal lobe* is the highest among mammals, including humans (Holdrege, 2001). These features of the elephant brain favor hearing, learning, memory, and emotion, all of which relate to communication and socialization. Similarly, elephants have an unusually large and convoluted *hippocampus* compared to primates and cetaceans. In humans, the *hippocampus* and formation of the *hippocampus* relate to recent memory of facts and events as well as emotions. In elephants this may also equate to social and chemical memory in which the identity of individuals is recorded and remembered.

Elephants can be right-handed or left-handed. When eating grass the elephant will grasp a clump of grass by the end of its trunk, preferentially insert the right or left tusk under the grass, and break the tuft by lifting and tearing with the tusk. Such persistent use wears a notch in the tip of the tusk, marking it as the preferred tusk.

In contrast to humans, the *occipital lobe*, composed of the *primary visual cortex* and *association cortex*, are small and ill-defined in elephants. These parts of the brain are related to a well-developed sense of vision. As mentioned earlier, an elephant's vision is one of its least-developed senses.

Finally, species with large average brains relative to their body size show greater ability to process and utilize complex information. We pursue this in the following chapters, showing and emphasizing in particular that the two senses discussed here, smell and hearing, play a major role in the functioning of an elephant.

In the above brief overview of the elephant's brain, there is evidence to support the role of smell, hearing, taste, and touch as important sources of input to the functioning of an elephant. Of the five recognized senses, the elephant's brain, in comparison to that of humans, is least well developed for sight. The prominence of smell and hearing over sight in the elephant's brain accords with the evolution of the species in dense, closed-canopy rain forests. It further stresses the fact that if we are to penetrate the mind of an elephant we must think in terms of smell and hearing and to a lesser degree in terms of sight. It is further likely that we will need to expand our conception of how olfactory and auditory signals as well as touch and taste are processed and used. In particular, it is likely that the elephant processes signals from multiple sources simultaneously and that an integrative rather than a reductionist viewpoint must be taken to penetrate elephant cognition.

# Chapter 4

# Functioning of the Brain

The brain of an elephant is a complex interactive organ that has been receiving, sending, interpreting, and storing millions of signals at rates exceeding many hundreds of pulses per second over millions of years (Eagleman, 2011). Much of the brain is a closed hard-wired system that runs internal functions such as breathing, digestion, and motor controls. This unconscious system operates without the intervention of the conscious but is not decoupled from conscious modulation. The unconscious is a repository for all of the preceding conscious lessons of survival. No brain of a newborn elephant is a blank slate. Instead, it is equipped by evolution to survive. The newly born calf will struggle to its feet within minutes of birth, it will find its legs and its mother's breast, and it will immediately imprint on its mother's voice and smell, shelter between its mother's legs, and follow her and the herd. Newly born zebra foals are believed to imprint upon the black-and-white striped patterns unique to its mother's hind quarters, allowing the calf to find its mother among dozens of other zebras within the herd. This survival instinct has evolved within the context of the elephant's immediate environment—its umwelt. Its umwelt includes not only the physical and living environment, but complex relationships of a highly social animal.

The senses of smell, hearing, touch, taste, and sight built by the conscious and stored in the unconscious are learned responses. We know from humans that the interpretation of the signals sent to the brain by these senses are in the form of electrical pulses. While the eye has complex receptors in the retina, it does not project an image on the brain analogous to that produced on the film in a camera. Instead, it generates a host of neural signals that must be interpreted by the brain as an image. The brain makes assumptions about what is seen. For example, for the human, light normally comes from above and shadows are cast below. The human brain automatically adjusts to that assumed distribution of light. Reverse this situation and all sorts of tricks can be played on the brain, including "seeing" water flowing uphill (see the lithograph of Dutch artist M.C. Escher titled *Waterfall* [1961]).

In the case of animals, we cannot be sure. While elephants use their eyes as we do, it is quite possible, given the relative role of sight and hearing for an elephant, that they may construct images from sensory systems other than or in addition to optical signals. In rare instances, such as Mike May who was blinded

Elephant Sense and Sensibility. http://dx.doi.org/10.1016/B978-0-12-802217-7.00004-1

at 3 years old but had his sight restored 40 years later, he was unable to construct images (Eagleman, 2011, p. 38). It took time for his brain to learn to see.

Hägstrum (2000) has suggested that homing pigeons use the earth's low-level seismic noise to navigate. Continual movement of the earth's crust produces a low-level seismic signal, which is recorded on a seismograph as a band of low-level noise. This low-level noise may be reflected and enhanced by the surface features of the earth, creating a map of the surrounding topography in the mind of the bird. Such a map of the bird's surroundings permits it to find the precise location of its dovecot. While it might use other means to determine the general location of its home (the stars, the earth's magnetic field), these guidance systems are insufficiently precise to pinpoint its ultimate destination. With the elephant's highly developed sense of sound, particularly at the very low frequencies of seismic noise, it is possible that the ability of the elephant to navigate and find specific locations is based on the natural seismic sound fields of the earth, which humans cannot hear.

There is further evidence (O'Connell-Rodwell et al., 2000, 2001, 2004, 2012) that elephants can project their powerful calls into the near-surface substrate of the earth, which then propagate as seismic signals over considerable distances. Elephants might be seen as standing on their "fingers and toes" embedded in a jelly-like base. Such a base is in excellent contact with the earth's surface.

Pacinian corpuscles in the toes and feet pick up transmitted seismic signals from other elephants and transfer these signals via the bone structure of the detecting animal to its auditory system. In this case, the seismic signal is converted by the elephant's brain with sufficient fidelity to allow the receiver to identify the sender (O'Connell-Rodwell et al., 2007). We remain unsure as to how the elephant recognizes the other individual. Is a visual image constructed? Is it simply voice recognition without an image? Is it converted to smell or all three? What we do know is that neural processes convert the seismic signal to recognition.

In both of the above cases, neural signals have been interpreted in ways unfamiliar to humans. Yet the possibility of sensory signals being converted to images by the brain has been demonstrated in humans. Humans can read braille through the touch of their fingers. Using a grid of 600 electrodes, Eric Weihenmayer is able to "see" using the sensitivity of touch of his tongue (Eagleman, 2011). In a condition called *synesthesia*, people can hear colors, taste shapes, and experience other sensory blendings. There is no obvious reason why animals with brains as complex as those of elephants cannot do the same thing. Certainly, there is evidence that cross-talk among sensory areas of the brain and visual and auditory systems are closely tied to each other. Eagleman (2011) refers to this as the advantage of a "loopy brain": a brain that is able to construct predictions of what will happen based on experience gained from the looping of circuits through multiple sensory and neural systems. Here perception reflects the active comparison of sensory inputs with internal expectations. Eagleman considers the human example of catching a fly ball in baseball. This involves accurately predicting complex physical processes: the striking of the ball and the sound of the bat against the ball, the trajectory of the ball involving velocities and accelerations (changes in speed, direction, and time), and programming the precise time and location of the simultaneous arrival of the ball and glove. None of this is done consciously but must be done almost instantaneously.[1] Yet ultimately it is based on learning and experience. So where does the unconscious begin and the conscious end? And if brains are able to orchestrate complex operations and rely mainly, if not entirely, upon the unconscious, might not animals with adequate brains in which to accumulate and store the unconscious processes be even better than humans in allowing their unconscious to function without the interference of the conscious? Anyone who has mastered a game to a reasonable level of competence has experienced the euphoric state of being in the "zone." You can do nothing wrong and play effortlessly at your highest level. Conversely, you cannot consciously repeat the performance and the harder you try the worse you perform.

The role of the conscious and that of the unconscious is exhibited by professional athletes all of the time. Those at the top of their respective games, while physically capable of performing at the highest level, cannot do so on a consistent basis. A dramatic example is seen in the struggle that Tiger Woods underwent to regain his status in the golf world. There is no doubt that he retained the physical skills to do so. What got in his way and prevented him from doing so is his conscious mind. In using its large brain might not an elephant be more effective in integrating all of the information being received from its sensors, call more effectively upon the stored information, recognizing what is important and what is not, and do so more effectively than a human? If survival is the measure of this process, then perhaps an elephant holds the edge over a human?

These questions revolve around many issues, not least of which is the implicit assumption above that elephants (and therefore other nonhuman animals) operate far more by the unconscious than the conscious. The immediate flaw in the argument is that the unconscious and conscious are not separated but are coupled. The degree to which this coupling can function may be as important as decoupling one function from the other. In the next chapters, we explore the ability of an elephant to modulate its considerable unconscious reserves with conscious behavior. Memory, we argue, is fundamental to the effective functioning of the mind.

---

1. The 2013 Nobel Prize in Medicine was awarded to three American neuroscientists, James Rothman, Randy Schekman, and Thomas Südhof, for their work in how cells or genes transfer information and substance.

# Chapter 5

# Memory

We, and elephants, are what we remember (Foer, 2011). Our very existence depends on memory. It is possible that everything learned is remembered, yet everything remembered cannot be recalled on demand. Even this statement must be qualified. In 2006, researchers at the University of California at Irvine reported on a woman, referred to only as AJ, who was capable of remembering in detail, with considerable accuracy, almost everything that happened in her past. They proposed the name *hyperthymestic* syndrome, from the Greek word *thymesis*, meaning "remembering," to describe the condition (Parker et al., 2006). Since then numerous cases have been recorded of people capable of almost total recall such that the ability to do so is called "superior autobiographical memory" (LePort et al., 2012). With this ability to recall manifesting in a small fraction of the human population, caution needs to be exercised on the possibility that such an ability can exist and be far more prevalent in another species.

Evolution has dictated that events critical to survival are remembered. Humans retain the ability to recall memorable events no longer because remembering them is critical to our survival but because our brains have been programmed to remember unusual or critical events. We can vividly recall the time, place, and what we were doing when we heard the news of President John F. Kennedy's assassination or of the attacks on the World Trade Center. Yet we cannot remember the name of a person we met just 5 min ago.

The mind of an elephant, as for all animals, has evolved to recall events critical to their survival. An exceptional memory is likely in a highly social animal of the size of an elephant. Continual contact and association with other elephants plus the neural demands of complex motor functions of a large-bodied animal demand a large brain and a good memory (Byrne and Bates, 2009).

There are three regions that are important to memory consolidation and are prominent in the elephant's brain (McGaugh, 2003). The *hippocampus* and the *medial temporal cortex* are important to the long-term consolidation of explicit memory in the cerebral cortex. Explicit memory, sometimes referred to as episodic memory, is memory of specific events. In contrast, semantic memory is derived from general knowledge. The *caudate* regulates body movements and responses to learning, which ultimately become automatic, such as knowledge of a frequently used game trail leading to a water hole. The *amygdala* plays a central role in the response to fear and as such may consolidate memories of traumatic events.

Elephant Sense and Sensibility. http://dx.doi.org/10.1016/B978-0-12-802217-7.00005-3

Memories are formed over time and can be seen to pass through stages. The consolidation of memories is not, however, sequentially linked but is based on independent processes operating in parallel. Lasting memories consolidate over time, even when initially triggered by traumatic events.

Elephant society is considered to be the most complex among all animals except that of humans. Moss (Moss, 1988; Moss and Lee, 2011, pp. 205–223) sees the family unit as the core of this social network. Extended families then form bond groups, clans, and ultimately entire populations (Figures 5.1–5.3). Leggett et al. (2011) and Wittemyer et al. (2005, 2009) have built upon and refined Moss's original social network in an attempt to include fission–fusion in the social organization to account for a fluid exchange of members and family units between the basic group and the wider populations. McComb and coworkers (McComb et al., 2000) have shown that individual elephants can recognize the calls of at least 14 other families, totaling as many as a hundred other individuals. Such a feat of individual identification would test the mental abilities of humans. Elephants keep track of family and bond group members on a continual basis. Low-frequency contact calls (rumbles) are emitted on a frequent if not near-continuous basis. These calls serve to maintain cohesion of the family group, defining the territory occupied and alerting other groups to their presence.

**FIGURE 5.1** Very powerful bonds are established during the long period of nearly 2 years of close contact between mother and calf. Females will remain within their family unit throughout their lifetime. Males will leave or be made to leave their family at puberty. They may well return briefly during their lifetime.

FIGURE 5.2   All members of the bond group will have close family ties. One elephant will recognize by sound, smell, touch, taste, and sight more than 100 other elephants within their home population.

FIGURE 5.3   Fluid movement, known as fission–fusion, between bond groups and populations exhibit a social network unique amongst terrestrial animals with the possible (past) exception of humans. The role and importance of this extensive bonding is unknown but potentially vital to the survival of this species.

While their vision in bright light is not good, perhaps limited beyond 50 m (160 ft), elephants detect and are very sensitive to subtle body language involving ears (flapping, raising, folding), trunk, tusks, feet, tail, and whole body.

Olfaction and chemical communication combined with the use of touch add to the complexity of memory functioning in a highly social system.

Experienced elephant matriarchs, carrying perhaps generations of knowledge passed on by successive predecessors, demonstrate higher rates of survivability in times of famine and drought (Foley et al., 2008) and higher reproductive rates within their herd (McComb et al., 2001) (Figure 5.4). The ability to transfer this knowledge to successive generations of elephants is a direct function of their longevity and strong social bonds. Female elephant offspring remain in the family unit throughout their lives. Future matriarchs of a herd are frequently the daughters of the herd matriarch (Figure 5.5). Knowledge embedded within the memory of the matriarch is transferred through experience over as much as 40 or 50 years. Long-term memory not only allows these elephants to recall remote locations but allows them to find these locations across considerable distances of featureless country (recall the introductory description of finding isolated oases in the Namib desert after not visiting these locations for more than 20 years).

Memory also serves to determine where and when specific food may be found and when it is edible. Cochrane (2003) has postulated that elephants adopt specific pathways that are committed to memory and lead to fruit-bearing trees, *Balamites wilsonia*, which are rich in proteins, fats, and diosgenin.

**FIGURE 5.4**   The matriarch leads the herd, drawing on a vast store of spatial memory that serves to preserve the herd.

FIGURE 5.5    A calf learns to identify mother and members of its family by sound, much of which may be below human hearing, touch, taste, and smell, with vision probably being last of the senses to be relied upon. All of this information is to be stored in short- and long-term memory.

The Mfuwe Lodge in the South Luangwa National Park in Zambia inadvertently enclosed wild mango trees in the interior atrium of the lodge. These mangoes ripen in November of each year, lasting for 4–6 weeks. A family unit of 10 elephants, led by their matriarch, Wonky Tusk, have returned to these mango trees from somewhere in the 9500 km$^2$ (3700 mile$^2$) national park in each November of the past 4 years. Wonky Tusk has led her group into the lodge, past the reception desk, to feed upon these mangoes without showing any aggression or fear of humans in the lodge (Carter, 2008; http://www.africatravelguide.com/articles/the-elephants-of-mfuwe-lodge.html).

Elephants can recognize other elephants as well as humans after being separated from each other for many years. Buckley (2009) at the Elephant Sanctuary in Hohenwald, Tennessee, reports that in 1999, Jenny, a resident elephant, was introduced to a newcomer, Shirley, an Asian elephant. Both elephants became animated, vocalizing loudly and using their trunks to check each other out. Neither exhibited any aggression, both displaying what observers described as a "euphoric emotional reunion." Probing the history of the two elephants, it turned out that Shirley and Jenny had been together in the Carson and Barnes Circus for a few months, 23 years earlier.

Elephants have highly sensitive olfactory and chemical detection capabilities. Within the nasal cavity of an elephant are seven turbinates (dogs have five), which are scrolls of bones with sensitive tissues to detect smell and which contain millions of receptor cells. In addition, elephants have a Jacobson's or vomeronasal organ on the roof of the mouth. This organ is highly sensitive to flemen, which is sampled by contact with the tip of the trunk and is then brought to the roof of the mouth. Sources of these odors are urine, feces, saliva, and the excretion of the temporal gland as well as the genitals found in both male and female African elephants. Adult African elephants of both sexes have a gland located at the temple which results in a dark stain marking the sides of their heads when agitated or stressed. Indian male elephants, but not females, show a similar glandular response, especially when in the heightened sexual state of musth. Elephants are seen to be touching and sampling these sources using the tip of their trunk.

Bates et al. (2008a) tested elephants' ability to keep track of their social companions using olfactory cues in urine. They found that in this way elephants can recognize up to 17 females and possibly 30 family members. Keeping track of this number of individuals is cognitively challenging and places considerable demands on the animal's memory capacity.

When the stable isotopes of oxygen, hydrogen, and nitrogen embedded in elephant tusks were related to the water holes of Etosha National Park (Dieudonne, 1998), the territories of the best matriarchs were found to be coincident with the best water sources.

It is likely that elephants have developed multiple sensory inputs of the kinds discussed earlier, including in particular a range of audible and inaudible (to humans) biotic and abiotic signals together with olfactory and chemical signals to produce exceptional spatial mapping skills. These mapping skills rely upon and interact with long-term memory skills vested in and passed on in successive generations by the matriarchs of the herd. Such abilities have direct survival consequences so that matriarchs occupy the best territories with respect to food and water and have the best survival rates under stress.

The evolution of an elephant's memory rests heavily upon spatial recall (Byrne et al., 2009). The location of food and water, the times of year when these are plentiful or scarce, and the likely presence of predators are all placed within the framework of a spatial map. The critical nature and the location of this knowledge make it memorable and thus consolidated within the brain. This ability to recognize critical facts and place these facts within a spatial context is programmed into elephants' and other animals' brains, including that of humans. Human memory contestants have clearly recognized this ability of the brain, developing the "Memory Palace" as a foundation for a system of remembering (Foer, 2011). Foer (2011) illustrates this capability by asking one to recall the house one was born and spent our time in as a child. If one has done so, recall of an amazing level of detail is possible. It is easy to visualize, after many decades with no intervening attempt to do so, minute detail of the entrance gate, the path to the front door, the steps and cracks in the steps up to the door, the screen door and handle, the entrance hall, and so on throughout the entire house. To remember things, in sequence, one can place what needs to be remembered, like a shopping list, at various locations in the house. The more memorable we make these placements, like imagining bananas sticking out of the mailbox, the more likely the item will be remembered. Images and not numbers or letters are what we are programmed to remember. The capacity to create a spatial neural framework has been burned over evolutionary time into the elephant's brain. Convergent evolution has apparently led to remarkably similar capabilities within the human brain, reflecting in that brain the same elements of storage and retrieval. The possibility that the elephant's brain is better equipped than the human brain to both store information and to retrieve it under these circumstances becomes entirely plausible.

To the author's knowledge, neurologists have not raised the question as to whether elephants or any other animal, other than humans, are known to suffer from memory loss similar to that observed in humans. If this is indeed so, it may be possible that insight into the workings of the brain may be furthered by inquiring whether and why such changes are not observed in a species such as elephants that depend as critically as we do upon their memory. And if not, why not.

In the chapters that follow we explore how the elephant uses its memory. Can it interpret the mental or social state of others? Can we detect the origins of morality in an elephant and has this morality evolved into feelings of empathy, emotion, and, ultimately, the emergence of intelligent behavior?

# Chapter 6

# Morality

Dawkins believes that animals at the very least exhibit "selfishness" and at the very best both "selfishness and altruism." He holds, however, that the primary manifestation is "selfishness" and that the origin of this behavior is the gene (Dawkins, 1989).

A significant conclusion that can be drawn from Dawkins's position and one that is central to this chapter is that he has no difficulty in ascribing an emotion, selfishness, to animals. He further accepts that animals may have other emotions that can include altruism and even morality.

It is more difficult to conclude what role, if any, Dawkins assigns to nurture and, in particular, to the role played by society in determining behavior in animals compared to humans. He is, however, careful to say in the *Selfish Gene* (pp. 3 and 4) that he is not entering the debate on nature versus nurture or providing a description of human and other species behavior. We will, on the other hand, deal explicitly with the role played by society in elephant behavior.

Dawkins seems to reject the notion that morality is based on evolution, arguing that while humans may be able to change the "ruthless selfishness" of the genes, other animals cannot. Holding that man as opposed to other animals is "uniquely defined by culture," he sees genetically inherited traits as being modified only by humans (*Selfish Gene*, p. 2).

In *Unweaving the Rainbow* (Dawkins, 2000, p. 212), Dawkins goes even further to say "much of animal nature is indeed altruistic, cooperative and even attended by benevolent subjective emotions, but this follows from, rather than contradicts, selfishness at the genetic level." Similarly (pp. 232–233), he accepts that co-evolution and co-adaptation between and within species can be positive and constructive while remaining fundamentally selfish but pragmatically cooperative.

While conceding the uniqueness of human culture, we explore the special nature of elephant society and the interplay between society and behavior.

In his book *Good Natured: The Origins of Right and Wrong in Humans and Other Animals*, de Waal (1996) concludes that evolutionary behavior crafted rules of survival that we and other animals have adopted and built upon. But he points out that only we still argue about the validity and purpose of these rules, failing to acknowledge that we did not formulate the basic concepts—nature did. And that "many of the sentiments and cognitive abilities underlying human

Elephant Sense and Sensibility. http://dx.doi.org/10.1016/B978-0-12-802217-7.00006-5

**27**

morality antedate the appearance of our species on this planet." Human morality derives from sympathy, attachments, succorance, emotional contagion, and cognitive empathy. He further points out that the human brain (as are all other brains) is a product of evolution and is fundamentally the same as the nervous system of other higher animals.

In *The Bonobo and the Atheist*, de Waal (2013, p. 162) defines morality "as a system of rules helping not hurting others where the community is placed before the individual. This does not deny self-interest but curbs its pursuit and promotes cooperation." Morality is a product of the species in which it evolves. It is grounded in emotional values that stem from social pressure and the role of authority. As we will see in the next chapter, the extended dependence of the young upon the mother and the group generates tolerance and a hierarchical structure. In the case of elephants, the ejection of the young male at puberty not only solves problems of disruption in the herd but allows females to remain as a cohesive group while simultaneously maintaining genetic diversity and survival characteristics. A single individual, the matriarch, with clear mental attributes, is elevated to a position of leadership extended over time, with no evidence of conflict or injury within the herd (Figure 6.1).

**FIGURE 6.1**   Bodily contact is continued into adulthood.

The morality of elephants is then grounded in this unique social system, which has generated over time not rationally constructed rules but behavior that has promoted survival of their species. It is these behaviors in which humans found irresistible challenges to elaborate upon and codify. As Philip Kitchers (de Waal, 2013, Bonobo, p. 171) aptly states, "Philosophers have cast themselves as enlightened replacements for religious teachers who previously pretended to insight." Morality is no longer solely the domain of philosophers.

Nevertheless, philosophers have a substantive and distinguished claim on the understanding and elucidation of morality. It is both essential and instructive to summarize the salient features of how modern-day philosophers view morality and in particular how they view the question of morality in nonhuman animals. In summarizing the current view of morality in humans and nonhumans, I draw mainly upon the recent book by B.A. Dixon, *Animals, Emotions and Morality: Marking the Boundary* (2008).

In any attempt to show that humans and animals are moral kin, expressed in the moral kinship hypothesis, Dixon places considerable emphasis upon the need to define clearly the concepts involved. If emotions form the basis for morality then it is essential that we understand what is meant by emotions. She recognizes four conditions that define emotions. Emotions, she says, must

- be directed at objects, persons, or the state of affairs;
- be interpreted by the agent recognizing that each agent may have a different experience or a different interpretation;
- involve beliefs, judgments, and appraisals; and
- involve a value judgment made by each individual.

Furthermore, morally laden emotions must be connected to moral virtues, which means that they must have cognitive, evaluative, intentional content. They must reflect good moral characteristics such as dignity and unwavering love. Compassion is cited as a morally laden virtue, which Dixon elaborates as reflecting empathy in action, is commendable, shows a sense of right and wrong, and is sensitive to context.

To illustrate the need to clearly define how emotions are coupled with morality, consider the following situation. A mother is watching her son climb a ladder to reach a high window. He slips and falls and seems to be injured. What are his mother's emotions and how does she react? Slightly alter the situation where instead of her son the mother observes what she interprets to be a suspicious-looking character climbing the same ladder to the same window. He also slips, falls, and seems to be injured. The mother's moral response in the two situations will likely be quite different. Each of the four above criteria that define the emotions needs to be considered before interpreting the moral response of the mother.

The logic of the conclusions drawn from Dixon's definition of emotions and connection between emotions and morality is that emotions are derived from conscious thought; morality in turn stems from emotions and thus from conscious thought. Since nonhuman animals are not believed by Dixon to be capable of conscious thought, nonhuman animals are not moral nor morally culpable.

Dixon rejects the concept that if closely related species act the same then the underlying mental processes are probably the same. Known as evolutionary parsimony, Dixon rejects this theory on the grounds that:

- it depends on how we define closeness between species and thus conclude that they exhibit similar moral behavior;
- if some animals exhibit moral emotions, then humans are not unique with respect to morality; and
- unless emotions are strictly defined, we cannot conclude anything from perceived similarities.

Dixon also rejects the concept of moral continuity and gradualism where the former is a continuum in cognitive complexity from a simple emphatic response to a complex mental comprehension of another's state, and the latter is where there is a gradual evolution from the simple to the complex. Likewise, Dixon does not accept de Waal's building blocks of morality thesis where morality can evolve from some primitive to some advanced level. In each case, Dixon's primary objection seems to be that the proponents of these theories fail to specify what is morally significant. Here Dixon insists that a response must be shown that is morally significant and that the subject has the capacity for empathy and is motivated to alleviate the object's distress.

Dixon has considerable difficulty in presenting a consistent argument that defines a boundary between children who should not be considered moral subjects and those who are. In particular, her rejection of gradualism and the building block thesis of evolving morality is seen to be violated by such statements. She then claims that because they are children "we aught to employ more liberal principles of responsibility" (p. 172). It is not clear how Dixon proposes to determine the threshold or boundary between children that are moral subjects and those that are not. Nor is it clear what the "level of responsibility" is that qualifies children to be viewed as moral subjects. Is it based on a single act that is judged to be moral? More than one act? How many? If a child does not suddenly cross a boundary from being judged not to be moral to being moral, does the child do this gradually and, if so, is this not gradualism that Dixon has rejected?

Only occasionally and only in the latter half of her book does Dixon concede that humans are animals. Elsewhere she makes the distinction between humans and animals. Her stance on whether animals have emotions or exhibit morality is not as rigid as a number of other current philosophers, such as Philip Kitcher who holds that animals cannot be considered moral subjects because they do not possess an "inner intellectual state," which they must exhibit if they are to be considered capable of morality. Or Richard Joyce because animals "do not have language" or Marc Hauser because he considers that animals "do not have self-awareness" (Dixon, 2008, p. 126). Dixon, nevertheless, does not agree with de Waal (2013), Bekoff (2002), Bekoff and Pierce (2009), Rowlands (2012), or Regan (2004), among others, who hold that animals exhibit morality and that morality predates the emergence of *Homo sapiens*. Once emotions and the emergence of morality is seen as ancient and evolutionary, much of the super-imposed complexity disappears and the relative role played by emotions and morality within and across species is clarified.

Work with very young preverbal human infants (6–10 months old and some younger) at the Yale Infant Cognition Laboratory by Wynn (2008) and her colleagues challenges the boundaries described above and supports the view that morality is rooted in long-term adaptive behavior evolving over immense time scales favoring the survival of social species.

Responses of 3- to 6-month-old infants were shown on the CBS program *60 Minutes* on 28 July 2013, to puppet behavior that was both "nice" and "mean." Seventy-five percent of the infants tested chose the "nice" puppets over the "bad" puppets, suggesting a recognition of right and wrong and a sense of justice. When the puppets behaved badly, 85% of the infants opted to punish the "bad guys." Positive and negative feelings on the part of the babies were evident, with the analysis showing that 87% of the subjects showed bias. Conclusions drawn by the researchers suggest that infants prefer those who are kind and who are like themselves. This results in a bias and response to punish others, creating a "them" and "us" situation that predisposes the infant to choose those who are more likely to help them survive. In published work (Hamlin et al., 2007; Wynn, 2008), the Yale researchers show that infants prefer individuals who help versus those who hinder others, indicating that they can assess individuals on the basis of their behavior toward others. This capacity, they conclude, "may serve as the foundation for moral thought and action." Wynn also shows that infants can detect intentional actions on the part of individuals and understand these as "good" or "bad" within their social context. Experiments with these infants suggest that it is unnecessary to establish any artificial boundary between those who cannot and those who can be perceived as moral subjects. Infants as early as 3 months old demonstrate an understanding of social entities and react to the mental state of others.

A clear evolution in judging and responding to social interactions is demonstrated as tests are applied to children of increasing age. By 9 or 10 years old, Wynn found learned behavior replacing the early inherent response. By the teenage years, education and learned culture displace the previous behavior, although elements of the instinctual response remained. These findings support the concept that morality evolves and accepts both the precepts of gradualism and of the building blocks of morality suggested by de Waal.

Before returning to consider morality within elephants, we must acknowledge the views of those who have considered whether and how humans should concede rights to nonhuman animals. Tom Regan (2004) makes the case for animal rights. He argues that even if we do not concede that animals possess sensory, cognitive, conative, and volitional capacities, our human moral sensibilities demand that we extend these precepts to animals in morally relevant ways.

Regan, with many others, holds that animals have a mental life; he calls it subjects-of-a-life, in which they experience physical pain, pleasure, fear, contentment, anger, loneliness, frustration, satisfaction, and a number of other emotions that we readily recognize. These animals, which he recognizes are above some primitive form of life, see, hear, desire, remember, anticipate, plan, intend, and may even know what matters to them. They should never be treated simply as resources for others to exploit. Treating animals without respect is morally wrong and if morally wrong, such treatment in the ultimate sense should be abolished. While such a view leads us into serious conflict with a broad range of socioeconomic issues such as the banning of killing of all animals for consumption, Regan proposes, at the very least, a principle of basic moral rights. This principle holds that all animals deserve the right to respectful treatment. Animals have certain basic moral rights and should never be treated as mere receptacles of intrinsic values and that they have the prima facie right not to be harmed.

In the extreme, a group of biologists have suggested that plants have electrical and chemical signaling systems, responses that resemble memory and exhibit behavior that in other species would be ascribed to a brain (Brenner et al., 2006). Despite rejection of such notions by most biologists and neuroscientists, the unexplained behavior of plants in apparently generating short- and long-term electrical and chemical neurotransmitter-like signals must give us some pause for thought when considering the existence of moral behavior in elephants.

In light of the above discussion, we seek in this section to uncover evidence of what might be characterized as moral behavior in elephants, building upon what we know of their cognitive memory and social organization.

Elephants are among the most social animals on the planet (Lee and Poole, 2011). Family units and bond groups led by the oldest female are stable and long-lasting. Leadership is rarely, if ever, contested and female rank and position is recognized without contention (Evans and Harris, 2008). There is fluid exchange, referred to as fission–fusion, among the core family unit and the wider bond or even clan groups. Bonding forces are strong and the ability to recognize large numbers of individuals (see above) is well established. Individuals, in recognizing and meeting other elephants they have not seen for some time, express their feelings with loud vocal greetings, much bodily contact, exploring each other's mouths, eyes, ears, and temporal glands with their trunks. These displays are not confined to individuals in the herd but are expressed by virtually all members of the group.

Elephant society promotes survival. While the group may spread out over a fairly wide area when feeding, near continuous low-level, low-frequency contact is maintained. The need to drink water arises on an almost daily basis, since adult elephants drink about 50 gal of water per day. To satisfy this need, matriarchs will emit a loud, low-frequency assembly call. The herd will gather together, shelter the youngest members in the line-of-file, and proceed to the waterhole in an orderly fashion. Predators such as lions who, particularly at the height of the dry season when water may be limited to only a few locations, frequent these locations would certainly take young elephants if not accompanied by the herd. In a herd the adults will join together and will intimidate and drive away even a large pride of lions (Figures 6.2 and 6.3).

At waterholes with limited water and access, a herd of elephants will show remarkable constraint, taking turns to drink and avoiding sullying the water

**FIGURE 6.2** Large male lions threaten young elephants while a group of lionesses sometimes can kill an adult elephant.

**FIGURE 6.3** A pack of hyenas represents a threat to calves and young elephants.

where they are drinking and using the overflow part of the waterhole to bathe and mud-bath. At a waterhole in what was then Wanki Game Reserve in Southern Rhodesia, a very large group of elephants numbering over 100 approached a very small artificial waterhole fed by a windmill-powered pump. The available water was contained in a shallow circular cement basin no more than 5 m (15 ft) in diameter. An outflow on the downstream side of the basin allowed excess water to flow down into a muddy area where it disappeared into the ground. The elephants were clearly stressed on an extremely hot late afternoon and were hurrying in a characteristic fast walk, clouds of dust rising from the path and obscuring the end of the line of animals. Watching the urgency with which the herd was approaching, one feared chaos at such a small watering hole, hopelessly inadequate to serve such a large number of desperately thirsty animals. Instead, as the leaders approached the small waterhole, they slowed down, dipped their trunks into the water, continued walking around the hole while pumping trunkfuls of water into their mouths, completing a 270-degree arc around the hole and exiting from the hole to either continue to drink from the muddy overflow or begin spraying water and mud over themselves from this location. This orderly behavior continued until the entire line of elephants had circled the cement pool and each elephant had had at least one turn to drink from the unsullied water.

The conduct of the herd suggests a recognition of ordered behavior, if not rules, which benefit the group as a whole. There is broad recognition of rank but this did not override the need to protect the young. The young were protected in the middle of the line-of-file at the expense of elders at the rear.

A distinguished team of scientists (Shannon et al., 2013) have recently completed a study in which, for the first time, is presented experimental evidence of the crucial role played by society in the lives of elephants.

Social skills are severely disrupted by the dislocation of elephant family units and translocation to unfamiliar territory. In a long-lived, closely knit, kin-based society such as in elephants, initial trauma may well be followed in successive generations by a loss of knowledge and a potential decline in neurological development.

Playback calls were broadcast by Shannon's team to an "undisturbed" population of elephants in Amboseli National Park and to a "disturbed" population in the Pilanesberg National Park. Whereas the Amboseli population (58 family groups) had been subjected to a minimum in human-induced disturbance with no traumatic events, the Pilanesberg population (16 family groups) had been subjected to extreme trauma and translocation. In the case of the Pilanesberg elephants, all of their older family members had been shot around them. As infants they had been captured amidst this chaos, tied to their dead or dying mothers, held initially confined to small cages, brought up among humans, translocated in closed vehicles, and ultimately released in an unfamiliar environment with no contact at any point to other adult elephants.

Female contact calls were broadcast to family units within both groups of elephants. In the first experiment, social knowledge was tested using three social categories of playback calls:

*Social category 1:* familiar, a call from a well-known individual within the family group's population.
*Social category 2:* unfamiliar, a call from a low-ranking individual within the family group's population.
*Social category 3:* alien, a call from a caller who was unknown to the family unit.

In the second experiment, the role of age and dominance was tested. Five individual contact calls were selected from each of the Amboseli and the Pilanesberg populations. Each of the five calls was then assigned to one of five age groups ranging from 15 to 55 years in 5-year intervals. Each call was controlled to represent the caller's age and dominance (see Shannon et al., 2013, p. 7). The results of these experiments are remarkably clear.

In the first experiment, the defensive bunching exhibited by the Amboseli elephants shows a clear escalation as a function of how well the caller is known to the group. The Pilanesberg elephants, on the other hand, failed to focus their defensive bunching in response to the most threatening individual.

In the second experiment, the Amboseli elephants were clearly able to assess the age-related dominance of the caller whereas the Pilanesberg elephants were unable to distinguish between the level of social threat presented by the older and younger callers. The behavior of the Pilanesberg elephants may have mainly reflected the absence of exposure to older, more experienced elephants. However, the results of the above study also reflect a much deeper-seated response to their initial trauma, resulting in neurological damage that impaired decision-making ability. In both cases, the effects of the presence or absence of a highly organized social structure are clearly manifest. Without a closely knit, kin-based society, elephant behavior fails to develop and transmit responses that promote survival.

Morality is intimately embedded within the raising of young and the strength of the social bonds that hold the wider family group together (de Waal, 2008; Peterson, 2011). There is a profound paradox between the drive for genetic self-advancement of the individual which is potentially at the expense and the survival of the group as a whole. In elephants the solution seems to be in the unique matrilineal structure of their society. The large, powerful, and potentially dangerous males are excluded from the group at puberty and are allowed to ultimately indulge in serious and sometimes fatal battles to promote their individual genes. This is at the cost of the males and not of the group. Genetic advancement is achieved while promoting the survival of the group.

The strong discipline and hierarchical structure that exists among male elephants is not imposed upon the group. When males are ejected from the herd at puberty, they form bachelor herds in which social learning takes the form of

**FIGURE 6.4**   A young male will be ejected from the herd at puberty and may join other young males in a bachelor herd or as a companion or askari to an older mature male. The askari learns from the older male and provides additional sensory protection to the older male.

competition for rank and dominance. Rules of behavior are learned the hard way but not at the expense of the herd as a whole. Young males, referred to as askaris, often join one or more mature males learning and benefiting from these experienced older males (Figure 6.4).

The existence of strict rules of conduct are clearly evident in the behavior of these young males. In a day-long following of a herd of elephants in the Etosha National Park in Namibia, young juvenile males in the herd were unusually rambunctious, chasing guinea fowl, screaming, and mock-charging members of the group and bushes. Just before going to drink at a waterhole a magnificent bull appeared, certainly measuring close to 4 m (14 ft) at the shoulder. As he slowly strode to the waterhole, all rough-housing among the young males ceased. The apparent leader of the gang of juveniles approached the large bull and knelt down on his forelegs, bowing to the incoming lord of the domain in a clear display of submission (Figure 6.5).

In another well-known incident, young male elephants who had been saved in culling operations in the Kruger National Park in South Africa were subsequently translocated to the Pilanesberg National Park northwest of Johannesburg. On reaching puberty these young males began a reign of terror in which they killed and tried to mate with rhinos, attacked tourists in cars, and killed one ranger. Their behavior approximated what in humans would be seen as that often shown by juvenile delinquents deprived of parental control. South African Parks authorities decided to move adult bulls into the Pilanesberg Park in the hope of restraining these rampaging youngsters. Within the space of a

**FIGURE 6.5**   Young male showing submissive behavior to an adult bull. (Pen and ink drawing by author.)

few weeks all abhorrent behavior had ceased and the older bulls had imposed order and authority, strongly suggesting that clear rules of behavior exist, that these rules are imposed upon juveniles by adults, and that without the presence of knowledgeable adults order does not exist.

An elephant calf will nurse from its mother for at least 22 months, after which the mother is likely to come into estrous again, mate, and conceive another calf. During this 2-year period the calf is almost in continuous contact with the mother. This contact, in multiple ways, involves all of the senses. Other kin, especially young adult aunts and sisters, will actively assist in caring for the calf to the extent that the term "allomothering" has been coined to describe their behavior. Often the allomother and other closely related kin will take direct care of the calf. Frequently, when leaving waterholes or feeding locations, the calf will ignore the matriarch's "let's-go" call and be left behind in a potentially vulnerable situation. The calf's caretakers will immediately notice the laggard, return to it, and shepherd the calf back to the herd.

A young calf at a waterhole with steep slippery banks will often be unable to get out of the waterhole (Figure 6.6). Almost all elephant watchers have seen mothers and others in the herd come to the aid of the youngster and haul it bodily out of the water and up the bank. In one instance the wet and muddy calf was too slippery and the bank too steep for its two aunts to grasp its body and pull it up the bank. After successive tries of standing on the bank failed, first one then both aunts got into the water and joined forces to push and lift the calf up the bank.

This too failed. One of the aunts then got behind the calf and, gently nudging it, got the calf to go parallel to the bank to a location where the bank had disappeared and the calf could walk out. Unfortunately, as the calf tried to walk out

**FIGURE 6.6** Two closely related kin, probably sister and aunt, with the mother standing by, about to assist a calf unable to get out of a steep-sided waterhole.

it sank right up to its belly in deep mud. The other aunt then got in front of the calf and with her legs, in a shuffling motion, created a channel through the mud that the calf could use to finally get out of the waterhole.

The aunts involved in this episode clearly recognized that the calf was in an unusual situation, that it was stressed, and that it was unable to find a solution. While there was no real danger in their helping the calf, they displayed clear empathy and altruism in going to the calf's assistance. The mother, on the other hand, simply stood by and observed the whole episode. Both aunts exhibited considerable ingenuity in solving the problem. The first and most obvious solution to simply help the calf up the bank did not work. This solution, common in many other similar situations, had probably worked for these two elephants in the past. In fact, it is not unusual to see the helpers use their tusks to modify the bank by reducing its steepness or even making footholds. When the direct or known solution failed, an alternate solution was devised. When that too seemed to fail the obstacle was recognized and dealt with, and after considerable effort the calf was extricated from the waterhole.

Such extended (timewise) and extensive (groupwise) care given over the years of dependency of the calf by its mother, allomothers, and other kin in the group form the foundation of elephant society. Within this social framework elephants not only display recognition and knowledge of others but display clear coalitions and alliances exhibiting reciprocity and cooperation. African elephants have an extensive vocal and gestural repertoire. They react to sound and smell and to extremely subtle body language of both their own and other species (including humans). They are aware of sounds and smells from both their biotic and abiotic environment.

These multiple means of communication operate over a prolonged period of time in which behavioral patterns are transferred from adults to juveniles. Behavioral patterns that promote survival of the herd are embedded within the females and passed to successive generations. Vested in the emotions that promote this survival a positive feedback loop is created where survival of the individual is promoted by this social system with the success of the system amplified through survival. Moral behavior through social norms contributes to the survival where sympathy, empathy, mutual aid, fairness, and conflict resolution work to promote survival becoming established as the moral underpinnings of the society.

Morality begins with the mother and her young. Rules governing what may or may not be done are first imposed by the mother and ultimately adopted by the community as a whole. The rules governing status or positions in the group establish rank and influence reciprocity, intentionality, and expectation. The dominance structure within an elephant herd provides guidance to resources and limits confrontation. To understand morality in elephants we need to examine the nature and occurrence of these elements of morality within an elephant society (Figure 6.7).

If morality is traced from its very beginnings there is both evidence that moral behavior promotes survival, especially in social animals, and that animals have moral emotions. Once the evolutionary roots of morality are granted, there is no subsequent going back and arguing that because humans have progressively added sophistication to the definition of morality, animals by these human designations are not moral. For example, philosophers may require that

**FIGURE 6.7**   Fights between mature males can be violent and prolonged battles, sometimes ending in serious and fatal injuries. Males in musth tend to win such fights and go on to become the consort of the female in estrous, staying with her and protecting her from the other males for 4 or 5 days.

morality develop artificial virtues such as justice and conclude that because animals cannot develop justice they are not moral beings.

Darwin connected social bonds to morality but does not believe social sentiments are sufficient to produce true morality. Ultimately, Darwin thinks that only humans are moral beings. de Waal (see Rowlands, 2012, p. 20) sees morality in evolutionary terms laying the foundations that humans ultimately built upon. However, all three (Darwin, de Waal, and Hume) take human morality as the benchmark and argue that animals fall short of this benchmark (Rowlands, pp. 22–23).

Others including Bekoff and Pierce (Rowlands, p. 23) do not set a benchmark and argue that animals act morally even though this may not be how humans behave. Bekoff and Pierce (Rowland, p. 23) see animal moral behavior within a social group as promoting well-being and limiting harm. They recognize three behavioral clusters as paramount in determining morality: a *cooperation cluster* including altruism, reciprocity, trust, punishment, and revenge; an *empathy cluster* including empathy, compassion, caring, helping, grieving, and consoling; and a *justice cluster* including a sense of fair play, sharing, desire for equity, expectations concerning descent and entitlement, indignation, retribution, and spite (Bekoff and Pierce, 2009, Wild Justice, p. 7).

Rowlands believes that this definition of morality is too broad and deficient in drawing clear distinctions between behavior and motivation. For example, the *cooperation cluster* consists mostly of behaviors while the *justice cluster* consists mainly of motivational states, agreeing with de Waal that what is important in morality is the underlying motivation rather than the actual behavior. So the reason why food is shared is more important than the fact that animals do share. Having said this, however, Rowlands fails to provide anywhere in his otherwise excellent discussion *Can Animals be Moral* a concise and clear definition of morality. The closest he comes (p. 8) is that only if an animal displays "concern" for another's welfare can you claim that it is exhibiting a moral response. Only later (pp. 32–36) does Rowlands provide clearer insight in discussing "morally laden emotions," which provide reasons for those actions such that animals can possibly be moral subjects if they are motivated by emotions that have a moral content. They are not, however, moral agents who are responsible for their actions. And ultimately (p. 254), "if animals can act for moral reasons then they are worthy of moral respect."

# Chapter 7

# Emotions

In our attempt to penetrate the mind of an elephant, we have examined some aspects of the structure of the elephant's brain focusing on those parts that relate, in particular, to memory and olfactory and auditory functions. We have stressed that compartmentalization of the brain must be considered in conjunction with the integrated functioning of the brain. More realistically, the brain operates as a highly interactive system with rapid and complex feedback between different parts of the brain that occur without conscious intervention. We have made frequent comparisons to human brains and human behavior and must conclude at this point that there are more similarities between cognitive functions of humans and elephants (and other animals, especially the primates) than there are differences.

The composition and content of the elephant's brain is to a large degree a consequence of evolution, so that what is referred to as nature goes a long way in determining who an elephant is. Nurture, on the other hand, composed of the environment in which the elephant lives, including how and by whom the elephant is raised, shapes what has been determined by nature. The powerful social systems with strong bonds and attachments link the social and emotional centers of the elephant's brain and there is a neural substrate to emotions and feelings. Elephants are who they are not just because they are born elephants but because they are nurtured in a given social system.

In this chapter emotions are described as the platform upon which morality is constructed. We are left with at least two difficult questions: Are animals motivated by emotion and, if so, do these emotions have moral context? Emotions that we as humans readily recognize such as compassion, sympathy, grief, and courage are all expressions of concern. Such concern can be both positive and negative. One can rejoice in the happiness of others or one could resent the fact that others are happy. The important point, however, is whether the emotion felt is focused on the welfare of others. If it is, this emotion reflects a moral response. So as Rowlands (2012, p. 8) points out, if animals display "concern" for another's welfare, that response can be seen as a moral response. Other difficulties are seen such as how does one recognize the existence of concern, where concern is a conceptual issue rather than an empirical issue, adding more evidence won't help. The issue,

Elephant Sense and Sensibility. http://dx.doi.org/10.1016/B978-0-12-802217-7.00007-7

according to Rowlands, is "whether animals can act on the basis of moral emotions or for moral reasons or even whether animals can be moral." These questions and others are pursued here:

*Can emotions with moral content be attributed to animals?*
*How do you identify emotions with moral content?*
*Are animals motivated by emotion?*
*Does emotion cause behavior?*
*Can animals be motivated by moral considerations?*

We also proceed in concert with major moral theory, as expressed by Rowlands (2012, pp. 72–75), that animals possess a broad array of moral rights and as such must be treated as "moral patients" where a moral patient is a "legitimate object of moral concern, i.e., it has interests that should be taken into consideration when decisions are made which concern or impact it" (Rowlands, 2012, p. 72). Furthermore, an animal is a "moral subject" when it acts for "moral reasons" but not a "moral agent" since it cannot be responsible for its actions (Rowlands, 2012, p. 82).

In *The Expressions of Emotions in Man and Animals*, Darwin (1897) grappled with the question of whether emotions could be recognized in nonhuman animals. Confining himself mainly to humans, Darwin was at pains to show that common ancestors had common habits, resulting in a connection across the animal kingdom. He thus believed emotions to be innate and inherited with recognizable body language nearly universal across species (Figure 7.1). In contrast to current thinking (Seyfarth and Cheney, 2003a; Soltis, 2010, 2013),

**FIGURE 7.1** The gestural repertoire of an elephant is large, expressed in terms of body language, chemical signals including glandular secretions, sound, and other subtleties that are not detected by humans. (Pen and ink drawing by author.)

Darwin made little or no connection between emotion and language or language and body language.

By examining the possible emotional content of elephants' vocalizations, Soltis (2013) has been able to assess a number of the above questions. Soltis bases his recognition of emotion in the vocal expression of elephants on two elements. The first is that physiological activity associated with emotional states (Levenson, 2003) can influence voice characteristics by neural intervention. Such intervention, for example, increases the tension of the vocal cords which in turn increases the frequency of the emitted sounds. The second is to view emotional intensity (high or low) and emotional quality (negative or positive) as each having one of these two characteristics (Mendoza and Ruys, 2001).

As described elsewhere, elephants produce strong behavioral and vocal reactions to both birth and death. Behavioral signs of emotion such as increased secretion of the temporal glands, urination, and motions of their trunks are accompanied by loud rumbles, trumpets, and screams. In the case of death or when inspecting the remains of a calf (skull, bones), low-frequency sounds, inaudible to humans, have been recorded and described as "low moaning sounds" when speeded up to levels that can be heard by humans. The emotional content of these sounds is extremely intense and few human listeners do not suffer a strong sympathetic emotional reaction to these sounds.

Elephants in a herd will emit loud calling following the mating of one of their group, called mating-pandemonium. The mated female will also emit extremely powerful and characteristic calls distinct from the pandemonium produced by her family. Collectively, these loud calls may attract the attention of other distant males (Moss and Lee, 2011, p. 120).

Elephants have been exposed to audio playback of disturbed bees in locations where beehives have been used to protect crops. The audio responses of elephants function as an alarm call. When these sounds made by elephants are played back to families, they elicit fear and flight even though no bees or sounds of bees are present. Soltis (2013) speculates that such simple emotional vocal output could have been the beginning of elephant communication systems. Alarm calls for different threats, requiring different behavioral responses such as alerting the herd to danger or the actual confronting of a threat, are all possible within the vocal repertoire of an elephant.

The emotional state of the elephant is reflected in a wide range of vocalizations that do not include alarm calls. As emphasized throughout the text, essentially no attempts have been made to record elephant vocalization on a continuous basis with the ability to detect both calls made and calls received. The conclusion drawn from recordings made from fixed microphones (the Mushara experiment; see Chapter 9) is that elephants carry on a low level of infrasonic calling, which is amplified by the detection of other calls. As technology improves it will be possible to record multiple exchanges between members of a herd on a continuous basis. Such records, together with other observations, will allow descriptions of how elephants express their emotions in their voices.

Emotions expressed in human terms and recognizable by humans are seen in orphan elephants raised at David Sheldrick's Wildlife Trust in Kenya (Sheldrick, 2012). These often very young elephants are brought to the shelter after varying degrees of traumatic experience. Vital to the success of the orphanage has been Dame Sheldrick's attempt to raise these elephants as much like elephants as she can. In the process, this has revealed through continual close contact most of the emotions we recognize in humans. These young elephants exhibit feelings that can be described as happy to the extent that handlers believe they can detect smiles. The calves certainly can have fun and indulge in play, help each other, and show compassion. Conversely, they can display envy and jealousy, be fiercely competitive, throw tantrums when they don't get their way, develop hang-ups, suffer from depression, miss one another, grieve deeply, and come close to what handlers describe as shedding tears. Most recently a newborn calf, rejected by its mother, was reported to cry, shedding tears for 5 h (http://metro.co.uk/2013/09/11).

Interpreting signs of depression, including sadness, is difficult in humans. While the similarity between the emotions of humans and those of elephants is pervasive, whether elephants actually shed tears and whether this is a sign of internal stress and depression remains unknown. That elephants can feel sad or depressed is widely accepted among those who have had close contact with them.

All animals detect and interpret signals transmitted consciously or unconsciously by others. This body language may be unmistakably obvious or it may be extremely subtle, involving more than one of the senses. The social structure of elephants, combined with their capacity to interpret multiple signals, would suggest that they are both highly sensitive to emotional signals from other elephants and to signals from other species, including humans. This capacity may have played a role in explaining the behavior of two herds of elephants on Thula Thula, the game reserve in KwaZulu Natal established by Lawrence Anthony. Anthony died unexpectedly on Friday, 7 March 2012, on a visit to Johannesburg some 640 km (400 mile) from Thula Thula. Two days later, on Sunday, 9 March, the first of the two herds of elephants in the reserve arrived at the main farmhouse. A day later the second herd arrived. Both herds remained in the vicinity of the farmhouse for about 2 days before disappearing back into the bush. A memorial service was held for Anthony on the next day, Thursday, 13 March. Neither herd had been seen near the farmhouse for the past 18 months. Anthony's son, Dylan, estimated that both herds had traveled for about 12 h each to reach the farmhouse.

It is likely that the scientific community would dismiss this story as an isolated incident inadequately documented; at best it would be considered as purely coincidental. In doing so, observations may be lost or ignored not because they are faulty but because no acceptable explanation is present. Explanations can be wrong; observations are invaluable and should be preserved at all costs.

Body language and the multiple ways in which emotions can be expressed and transmitted may offer a plausible explanation to the response of the two elephant herds on Thula Thula to the death of Lawrence Anthony. News of Anthony's death would have reached Thula Thula within minutes of the event on Friday. Dismay, translated to grief, would have spread rapidly amongst the people on the farm. Women would have expressed their grief by loud ulutation. From that point onward for the next number of days, activities of an unusual nature would have occurred among the human population in an increasing fashion. This heightened level of activity and the change in the level of activity from the norm would have been easily detected by both herds of elephants. It is within reason to speculate that the character or nature of the activity reflected the emotional state of the people and that this was translated by the elephants into evidence of grief and sorrow. Whether elephants could make such an interpretation is dealt with elsewhere in this book. Whether they made such an interpretation in this case may only add a contributing factor to the accumulating evidence that they could detect that something was wrong. This alone may have triggered their pilgrimage to the farmhouse. Equally, we could argue that through these mechanisms both herds had in fact made what can only be seen as a sophisticated interpretation of the observed activity and did in fact go to the farmhouse to express grief and compassion.

# Chapter 8

# Empathy and Altruism

Empathy is the capacity to

- be affected by and share the emotional state of another,
- have some concept of the reasons for the other's state, and
- be able to sense what the other is feeling.

Empathy is a mammalian trait and almost certainly originates with parental care. It is essential in highly social animals such as elephants (Byrne and Bates, 2010). In elephants, as for mammals as a whole, maternal care is the costliest, longest-lasting act of all. Elephant young nurse for at least 22 months, remain close to their mothers for more than 10 years, and, if female, may remain with the mother's herd for the rest of their lives. The capacity to relate to the feelings of others is not only consolidated in the individual's maternal relationship but the process itself has been repeated in countless generations during evolutionary time. Empathy is for an elephant no conscious cognitive skill but manifested largely as an unconscious automatic response. It becomes part of its emotional response system and, if anything, displays empathy as an inborn emotional state.

Lee and Moss (2012) carried out a carefully controlled experiment to determine whether stable characteristics of elephant personalities could be traced from generation to generation. Four components of behavior—leadership, playfulness, gentleness, and constancy—were tested and the results subjected to statistical verification. The results showed that elephants are highly affiliative and cooperative and display infrequent overt aggression between family members. Care provided to calves made up most such interactions, followed by female–female friendly contact. Leadership is shown in exerting influence rather than dominance. Lee and Moss concluded that leadership was manifest in terms of respect, which recognized problem solving and permissiveness.

Elephants are very sensitive to the emotional state of others and have multiple and subtle means (sound, smell, sight, taste, and touch) of detecting this state. The capacity of the elephants to detect and read the state of others is critical to their survival. It is essential in social interactions, in coordinated activity, and in cooperation toward shared goals (de Waal, 2008). As Bates et al. (2008b) point out, not to do so would be maladaptive and potentially fatal.

Altruism is behavior that costs the participant something while benefiting another. Altruism grows out of empathy for those in need, blurring the lines between self and others and between selfish and unselfish (de Waal, 2013, pp. 27–28, 33).

Paolo Torchio photographed a family of some nine elephants surrounding and protecting one of their group giving birth on the open plains below Kilimanjaro in the Amboseli National Park (Steyn du Toit, http://www.2oceansvibe.com/2012/03/01/elephants-protect-female-giving-birth-from-prowling-predators). Hyenas were circling the group, aware of what was taking place and posing a distinct threat to the mother and calf. The tight circle around the mother was maintained until the calf was born.

Under other circumstances in the absence of predators and with cover available, the expectant mother more frequently leaves her group and goes into the bush to give birth. In the open plains of Amboseli, this was not possible and the herd rallied to give shelter.

At and following the moment of birth, Torchio recorded that "the elephants started trumpeting as though they were welcoming the new arrival." Moments later they tusked up dirt and grass, throwing clumps into the air. There is little doubt that the members of this group were aware of the physical state of the mother, knew that she was about to give birth, knew that predators were present, and represented a threat. Part of their perception was the clear recognition of her condition, but a significant part anticipated what had not yet happened. The elephants in the group may have sensed how the mother was feeling. Their display of tossing soil and grass into the air once the calf was born could have represented an expression of relief and joy or perhaps a physical response to mask the smell of blood and placenta from the nearby predators. Each of these possibilities represents the further capacity to perceive and respond to the state of others.

de Waal (2008) describes a blind elephant being cared for by an unrelated female. The blind female depended entirely upon the other, staying close to her and vocalizing as soon as they lost such contact. Both elephants appeared to be conscious of the fact that they depended on each other (p. 53).

Bates and colleagues (2008b) have carried out the first carefully controlled experiment on empathy on 58 family units in the Amboseli population of some 1434 elephants. They found 249 empathic responses, which were observed on at least two occasions. These they divided into six categories and attempted to determine the minimum level of cognitive activity (thinking) associated with each category.

They were able to show coalitions between two or more elephants who act against one or more other elephants who might threaten them or another elephant (calf) whom they may be protecting.

Acting collaboratively against another was observed and was taken as evidence of empathy and the recognition of the emotional state of others.

Protection was clearly observed, especially in mothers and allomothers who recognized the condition of others (potential threat) and acted to avoid the situation before it could occur such as protection against predators, intervening in play fights that were getting out of hand, pushing away a harasser, and avoiding others that were potentially dangerous or avoiding dangerous areas. Such pre-emptive behavior suggests both empathy and cognition.

Mothers and allomothers were found to give comfort, responding to the emotional state of the calf or anticipating that a stressful state was developing. Although the Bates study apparently did not include sound as part of their observations, it is likely that near continuous low-level, low-frequency calls play a pervasive role in comforting the young and reducing stress. Such comforting included care provided by others in the calf's natal family, behavior observed in only a few other species. As described above, calves often stray from their mothers and become separated. On occasions other females, not in the family group, will actively attempt to kidnap the calf. Calves will sometimes but not always give lost calls, which will be responded to by their mothers or allomothers. In many cases, retrievals will be initiated by a retriever who is not the mother of the lost calf, once more suggesting empathy and reasoning.

Help given to a calf in difficulty, as illustrated in the calf trapped in a waterhole described in chapter 6, is often carried out by elephants other than the mother, reaffirming the cognitive as opposed to parental instinctual aspect of the behavior. Both mothers and close kin have been seen breaking the wire of an electric fence with a tusk to allow the calf to get through the fence, leading the calf to easier terrain, or tusking a bank to give the calf a foothold and helping a calf stand.

The recognition and removal of foreign objects was also recorded in the Amboseli study and reported in a separate earlier study (Bates et al., 2007). An adult male removed a tranquilizing dart from another elephant. The dart was then dropped and ignored, suggesting that the motivation came from awareness that the dart was a foreign object and did not belong in the body of an elephant. Similarly, Maasai spears were examined. The elephant touched the spear, splashed water and mud onto the wound, then dusted the wound but did not remove the spear. While no evidence could be found, it is likely that the spear was not only visibly foreign but could have been clearly recognized through smell as alien and belonging to a species known to be dangerous (see Chapter 11).

The Amboseli study supports the contention that elephants recognize emotions in others of their species, are aware of the characteristic behavior and therefore react to unusual conditions, and understand that other elephants are animate agents who perform directed behavior and experience recognizable emotions. This cognitive empathy, however, must rest upon the ability to distinguish between self and others and to comprehend that others have selves like themselves.

Elephants' awareness of others or, at the very least, an acute awareness of their surroundings is illustrated by an incident observed by Jennifer Dieudonne in the Etosha National Park in Namibia in 2010. Dew drops tracing the intricate geometry of a giant spider web stretched across a game path were glinting in the rays of the rising sun. A herd of elephants was coming down the trail heading straight toward the jeweled web. Just before reaching it, each giant animal, including the younger calves, stepped off the trail and carefully went around the web.

The so-called "mirror test" provides a narrow test of self-awareness (Bates et al., 2008b). Perhaps only three animals—humans, apes, and dolphins— understand that they see themselves in mirrors. In human children, cognitive empathy and mirror recognition occur at the same time. Elephants have only recently been shown to recognize themselves in a mirror. Here the study was conducted by first allowing an elephant to become familiar with a mirror, then placing a mark (a prominent white X) on the head of the elephant, visible only to the elephant when it looked in the mirror. The elephant being tested touched the white mark with the tip of its trunk, clearly recognizing that this symbol was not part of its body. This evidence of self-awareness is considered essential for animals to exhibit empathy. Empathy, with the awareness of how others feel, can hardly be recognized without being aware of one's own feelings.

Other mirror tests with elephants have failed. Does this inconsistent result demonstrate that elephants are not aware of themselves and therefore not aware of the emotional state of others, and thus not capable of empathy? It is more likely that the deficiency lies in the design of the mirror test and the conclusions drawn from these tests rather than the absence of empathy in elephants. Sight in an elephant, as emphasized throughout this book, is not at the top of the list of sensory inputs to an elephant's brain. Smell and sound are more important to an elephant than sight. Touch and taste may well play a much larger role than sight. Thus, a test based on vision and the interpretation of images may not be the most appropriate test for elephants. The fact that an elephant, as opposed to a primate or a human child, accepts an image in a mirror as being of itself is remarkable. The elephant when recognizing others does so most directly in terms of smell, touch, feel, and sound rather than sight. For a mirror test to work for an elephant on the basis of sight alone is far more remarkable than for the same test to work for animals such as ourselves or other primates. de Waal (2008, pp. 116–117) adds percipient insight into this problem by saying, "We may be giving the wrong tools or holding up the wrong mirror." This insight is reflected in the famous dictum of experimental psychology that "absence of evidence is not evidence of absence."

The reaction of elephants to distressed and dying individuals and to the death of another elephant, especially one closely related to a given individual, may be the ultimate test of whether elephants exhibit empathy or not. The recognition of death or even the response to distress and dying is rare in most animals. Without the recognition of self, death of others has little or no significance. Lack of the impact of death and dying may therefore reflect profoundly upon the capacity for empathy.

There are many examples of elephants trying to assist those in distress or dying. Agitation is seen in the herd and herd members try desperately to lift the victim with their trunks and tusks. This has been observed both with natural mortality as well as when elephants have been immobilized by darting. A case in point is Echo's calf in Amboseli, who, at birth, was unable to straighten his front feet, struggled to get up, and then hobbled on his knees. He was repeatedly helped by other elephants and catered to by his mother, who remained with him even when the herd moved on (Moss, NGS film).

Douglas-Hamilton and colleagues (2006) were able to document the collapse and death of Eleanor, the matriarch of an elephant family (First Ladies) and the behavior of members of other families known to Eleanor. These elephants in the Samburu region of Kenya had been monitored by GPS tracking technology over a period of almost 10 years. Family units and their relationships were well known, as were the relationships between the 12 family units in the region.

Eleanor was observed by a member of Douglas-Hamilton's team to collapse from serious injuries sustained in a recent fall. Within 2 min, Grace, the matriarch of the Virtues family, arrived and bodily lifted Eleanor back onto her feet, trying to get her to walk. Weak and wobbly, Eleanor's back legs gave way and she again fell to the ground. Grace, trumpeting and very stressed, tried without success to get Eleanor onto her feet. At this point with night falling, Grace's family left, leaving Grace with Eleanor for at least the next hour.

Eleanor died the next day at 11:00 A.M., leaving her 6-month-old female calf confused and hungry. A total of 12 female members of Eleanor's family, in particular Maya, who was thought to be Eleanor's daughter, and members of five other family units could be tracked and their location relative to Eleanor's body determined. Over the next 6 days Douglas-Hamilton's team was able to monitor the movements of these other elephants relative to Eleanor's body. Family members approached the body more closely and spent more time with the body than nonfamily members, exhibiting altruistic behavior to kin. However, the interest displayed in the injured and subsequently dead elephant irrespective of genetic relationship showed a generalized response to distress not restricted to close kin. They concluded that this case could be seen as an example of "how elephants and humans may share emotions, such as compassion, and have an awareness and interest about death" (Douglas-Hamilton et al., 2006, p. 15).

In Etosha National Park, anthrax is endemic and can be fatal to elephants. Ginger and Conrad Brain, in the National Geographic film *Giants of Etosha*, recount how the matriarch of the herd they were following, Knobnose, lost two successive calves to anthrax. Knobnose, who had led this herd for many years, disappeared for months. Finally, she was located and could only be described by Ginger as extremely despondent and depressed. She would not lead the herd and was found visiting the remains of her calf. While gently touching and probing the skull of her calf with her trunk, Ginger recorded, inaudible to the human ear, the sounds Knobnose was making. When the recordings of these sounds were sped up and made audible to human ears, one heard the heart rendering moans of a mother grieving for her lost child. By any measure, Knobnose was expressing grief and was deeply aware of who it was that she was touching and holding in her trunk. Not only was her behavior clear evidence of grief, but it also showed that she perceived the depth of the relationship between who her calf was and herself. Knobnose finally mated again, bore a healthy calf, regained her equilibrium, and returned to lead her herd once again.

Numerous observers, including Darwin, believe that under extreme conditions of stress, elephants can break down and weep, shedding tears and uttering cries (see also Chapter 7).

Elephants visit the remains of their kind and have been shown (McComb et al., 2001) to recognize the remains of elephants among the bones of other animals. In contrast to the account above, McComb found that there seemed to be no particular recognition of kin, although when tusks were among the remains, these drew special attention. She also found that elephants can distinguish the bones of other elephants from those of large mammals and other nonelephant remains.

Joyce Poole describes an incident involving Eleanor, one of the orphans raised by Daphne Sheldrick. A woman wearing an ivory bracelet was warned to hide the bracelet behind her back. She did so but Eleanor reached around behind her, took her hand in her trunk, and raised it up close to her eye to look at it.

van Graan (cited in *The Game Rangers* Jan Roderigues, p. 56, 1992, permission granted by Roderigues), in dealing with crop-raiding elephants on the border of the Kruger National Park, had to shoot one of the bulls. The carcass of this bull was dismembered for later use. However, as night descended, the feet, tusks, trunk, and upper ear were moved and placed on an embankment some distance from the carcass.

On return, shortly after sunrise the next day, there was no sign of the dismembered parts. Numerous tracks of elephants did, however, mark the spot where the body parts were left. A search for these parts found them all back in a neat pile beside the carcass of the elephant. They, and the carcass, were partly covered with soil and vegetation. There were furrows in the soil where the ground had been disturbed with clear prints of the feet of the elephants who had done the work. Once again, there is clear recognition of who the dead elephant is, that his being has been disturbed and what belongs to him violated. It is more challenging to speculate on what motivated the other elephants to cover the body and the dismembered body parts. Reports of such behavior, however, are not uncommon (see, for example, the case of birth earlier in this chapter).

It is difficult not to conclude that elephants are aware of the distress of others, empathizing with vocalization and other responses to their distress and attempting to alleviate the suffering of others or restore others to a better state. When this fails, they identify with the remains and may even treat them with what could be called respect.

Elephants certainly recognize suffering and stress in other elephants and respond in ways that attempt to relieve such conditions. This response is not restricted to kin but is a generalized response to suffering and death of conspe-cifics. That elephants should show such a response is almost certainly related to their highly social structure. The death of the matriarch Eleanor led immedi-ately to the death of her calf and the loss of her knowledge to that family unit. In a more generalized sense, altruistic behavior exhibited to kin promotes the survival of the group. In evolutionary terms, natural selection of a beneficial trait such as empathy or altruism within the gene pool of the family benefits survival (Douglas-Hamilton et al., 2006; McComb et al., 2001).

More rarely elephants have been reported to respond to other species includ-ing humans in distress. A nearly blind elderly woman returning to her village was overtaken by nightfall. She was too frail to climb into a tree and resorted to sitting with her back against the trunk of a large tree. Elephants arrived at this spot during the night and, finding the woman, covered her with branches and vegetation and remained around her for the duration of the night.

More recently, Hutto (2014), in *Touching the Wild: Living with the Mule Deer of Deadman Gulch*, reports what he believes to be clear recognition by a mule deer of the death of her fawn and even the deaths of others within a much wider group of mule deer. Observing behavior of animals requires, as is demon-strated by Hutto, both in the case of wild turkeys (Hutto, 2006) and mule deer (Hutto, 2014), a significant commitment of effort and time.

The effort by Bates and her colleagues to examine elephant empathy under controlled conditions in the wild points the way to how a significant part of the research required to understand elephant cognition must be done. Failure to place the elephant in its natural environment is likely to seriously bias if not invalidate most experiments. The role played by elephant society and the full spectrum of the environment prohibits acceptable design of experiments under most artificial circumstances.

Experiments such as the mirror recognition test may be as invalid for elephants as they would be for bats. We as scientists are further heavily influ-enced by the need for rigor, recognizing the weakness of findings that cannot be quantified and our tendency to believe in the reductionist approach. Elephants, in particular, are likely to depend on multiple sensory inputs, are able to assimi-late such multiple inputs, and deal with feedback between them resulting in a response difficult to comprehend by humans that think primarily in terms of a single sensory system: sight.

As a cautionary note to an earlier remark, Michael Finkel, in the July 2013 issue of *National Geographic* magazine, reported on Daniel Kirk who had lost both eyes to retinal cancer at the age of 13 months. Daniel, at the age of 47, has taught himself to navigate by echo location. He produces clicking sounds, sometimes as fast as twice per second, which not only allow him to create im-ages in his mind of objects as much as 30 m (100 ft) to 45 m (150 ft) away, but to ride a bicycle on a city street. Close to 1000 blind students in over 30 countries have been taught to use echolocation.

# Chapter 9

# Communication

Communication, especially in a highly social animal such as an elephant, may have played a significant role in their evolution and in natural selection. Animal communication favors callers whose vocalizations benefit their listeners (Seyfarth and Cheney, 2003b). As is described here, elephants vocalize most often in the presence of other elephants, emphasizing the social function of communication. Although signalers may vocalize to change a listener's behavior, there is no general acceptance by researchers that animals call to inform others. The concensus amongst psychologists is that listeners acquire information from signals mainly by eavesdropping and not because the signaler intends to provide such information (Seyfarth and Cheney, 2003b). Dawkins (1989, p. 57) and Dawkins and Krebs (1978) believe that all animal communication represents manipulation of the signal-receiver by the signal-sender.

We show here and elsewhere that observed elephant behavior would suggest that such a view may not be entirely valid. Similarly, there is fairly universal agreement that the cognitive limitations of animals are responsible for the perceived differences in animal communication and human language. While the reach and extent of human language goes far beyond that of elephants, elements of the understanding of an elephant receiving a signal of the signaler's intent are becoming increasingly apparent. The ability of elephants to recognize the mental states of other elephants and to vocalize with the specific intent of informing others and transmitting to those listeners knowledge that the signaler possesses may equally well exist.

Individual recognition is central to social life. Elephants have knowledge of members that are within their own family group as well as in the wider population. This recognition, however, is primarily in the form of sound and smell and not sight. McComb et al. (2003) show that adult female elephants are familiar with and know the vocal identity of 14 families within the population, totaling some 100 individuals. While the full range of low-frequency calling patterns of elephants in the wild is unknown, it is likely that adult females maintain near-continuous contact via these calls. Older females in the herd have the most extensive knowledge of the calls and identities of other elephants as well as sounds emanating from other sources. Processing and storage of this range of auditory input requires considerable cognitive ability.

Elephant Sense and Sensibility. http://dx.doi.org/10.1016/B978-0-12-802217-7.00009-0

Elephants generate and can detect sound over the widest range of frequencies of all mammals (see http://people.eku.edu/ritchisong/RITCHISO/infrasounddiagram.gif). The female Asian elephant tested by Heffner and Heffner (1982, 1984) was able to detect a 60 dB signal as low as 17 Hz and as high as 10.5 kHz, which is the widest range known for any nonhuman mammal tested. In comparison, the range of human hearing is 20 Hz to 20 kHz (Soltis, 2010).

## SOUND GENERATION

Sounds generated by vertebrates depend on lung capacity and the mass, length, and elasticity of the vocal folds in the larynx. These fundamental sounds are then modulated as they pass through and emerge from the passageways that constitute the vocal tract. The vocal tract acts as a filter and operates independently of the source (Fitch and Hauser, 2002).

This independence between the "source" and the "filter" is considered to be the best current working hypothesis of how animals produce sound (Fitch and Hauser, 2002). However, much of what is known about the relationship between the source and filter has been learned from the study of humans, nonhuman primates, and other animals such as species of deer (Fitch, 2000; Fitch and Hauser, 2002; Fitch and Reby, 2001; Reby and McComb, 2003; Titze, 1994; Willmer et al., 2000; Wilson et al., 2001). Little is known of the source-filter relationship in elephants (Reby and McComb, 2003), yet it is useful to view the production of low-frequency elephant calls in these terms (Garstang, 2004).

Air driven from the lungs sets the vocal folds in the larynx in motion. With their own elasticity and mass, these folds, responding to the air flow over them, act as mechanical vibrators that can generate self-oscillation (Fitch and Hauser, 2002; Titze, 1994). When the folds close to the appropriate "phonatory" position, they generate acoustic energy. The period and thus the frequency of the opening and closing of the vocal folds produces the fundamental frequency (FO). This frequency is set passively by muscle tension, mass of the vocal cords, and lung pressure. There are small nonlinear oscillations around this fundamental frequency. These nonlinearities in the periodic vocal production provide structure to the morphology of the call and have been described in terms of deterministic chaos (Reby and McComb, 2003). Because the length, mass, and elasticity of the vocal folds are related to body size, the FO can be related to body size. However, these parameters (length, mass, elasticity) can change (e.g., with age) and the relationship between FO and body size is not robust (Fitch, 1997; McComb, 1991; Reby and McComb, 2003; Rendall, 1996; Riede and Fitch, 1999). The inverse relationship between the length and mass of an elephant's vocal folds predicts that it is capable of producing lower-frequency sounds than any other terrestrial animal. The prediction that larger animals produce lower frequencies than smaller animals of the same species has not been well verified by observations (McComb, 1991; Reby and McComb, 2003).

The supralaryngeal vocal tract of the elephant is the respiratory tract from the larynx to the tip of the trunk. For the elephant the pharyngeal pouch, nasal cavity, membrane near the tip of the trunk, the highly mobile tip of the trunk and the length and ability to change the length of the trunk are, in combination, unique Elephantidae features (Garstang, 2004) (Figures 9.1 and 9.2). Their individual and collective role in controlling the air column in the vocal tract has not been carefully studied.

Recent work, however, at the University of Vienna on an excised larynx of a 25-year-old female African elephant has both confirmed earlier supposition and described a number of new phenomena within the elephant vocal anatomy including the generation of vibrations of the vestibular folds, which increased the sound pressure levels by 12 dB. The study also showed that the anatomy of the elephant's larynx is more complex than that of a human and is capable of producing multiple wave patterns (traveling, standing, and irregular vocal fold vibrations) (Herbst et al., 2013).

The air column in the vocal tract has elasticity and mass that will vibrate preferentially at certain frequencies termed *normal modes* or *resonances*. In the simplest terms this column of air acts as though it is in a tube closed at one end such that the length of the tube is one-quarter of the wavelength. Wavelength is directly related to the speed of sound and inversely related to

**FIGURE 9.1**   The vocal tract of an adult elephant including the trunk, the nasal cavity seen as a bump on the forehead, the pharyngeal cavity, and the larynx can amount to a length of nearly 4.4 m (15 ft). Both the exterior trunk and the larynx can be extended by another foot.

**FIGURE 9.2**    A cavity in the throat of the elephant called the pharyngeal pouch can be filled with water (up to 4 or 5 L) and used as an emergency drinking or cooling supply of water and influence the sound emerging from the vocal tract of the animal. (Pen and ink watercolor on wood by author.)

the frequency. Thus, for the average speed of sound in the atmosphere near the ground of 350 m (1148 ft) per second and a frequency of 20 Hz, the wavelength is 350/20 m = 17.5 m (57 ft), and the vocal tract length would be (1/4)(17.5 m) or 4.4 m (15 ft). As we have seen, this is a realistic length of the vocal tract for an adult African elephant.

The vocal tract will act as a filter depending on the transit time of the sound waves up (and down) the column. The speed of sound in this tract will govern the transit time and thus will depend on the composition of gases in the tube and the temperature of those gases (Pierce, 1981).

The filter will act upon *all* of the frequencies being generated in the larynx (i.e., the FO and the nonlinear oscillations about the FO). The filter will thus shape the final form of the vocal signal. Prominent in this ultimate form of the signal will be "formants." These formants are selectively amplified parts of the vocal signal, clearly visible in the sonogram of the call of an elephant as nearly equally spaced bands of acoustic energy (Figure 9.3). Embedded in this signal envelope but independent of the formants is the FO and the integer harmonics of the FO.

The vocal tract length governs formant spacing. Formant spacing is a better predictor to body size than the size of the vocal folds or larynx (Fitch, 2000; Fitch and Hauser, 2002; Reby and McComb, 2003). In addition to changing the length of the vocal tract by extending its trunk, the elephant may be able to elongate the vocal tract by contraction of the larynx or laryngeal descent (Fitch and Reby, 2001; Reby and McComb, 2003). The presence or absence of water in the pharyngeal pouch may also influence vocal tract length. The function of the narrow connective strip of tissue dividing the lower part of the trunk into two

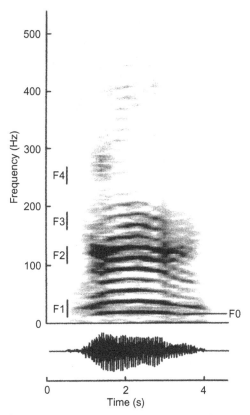

**FIGURE 9.3** Waveform of a female contact call showing the fundamental frequency (F0) and harmonics and the position of the first four formants (F1–F4). *Based on Garstang (2004); from McComb et al. (2003).*

orifices extending 10–13 cm (4–5 in.) above the nostril openings and the ability to open or constrict the nostril openings is unknown.

The elephant is a very large mammal with a unique vocal tract in the form of an elongated proboscis or trunk. The lung weight of an African elephant is about 22 kg (48.4 lbs), the vocal folds measure around 7.5 cm (3 in.), and the total length of the vocal tract measures between 3 and 4 m (9 and 13 ft). The elephant can extend this vocal tract by as much as one-fifth (20%) and can produce sounds as loud as 117 decibels (dB), where 120 dB is the level beyond which human hearing might be damaged (Soltis, 2010).

## SOUND DETECTION

The hearing mechanism of the elephant is also large. Each outer ear or pinna measures on the order of $1.8 \times 1.1$ m ($5.76 \times 3.52$ ft or more than 20 ft$^2$) and is highly mobile so that it can be extended outward and raised (together with the

**FIGURE 9.4**    Raised head with fully spread ears. (Pen and ink drawing by author.)

head), indicative of listening (Figure 9.4) (Moss, 1988; Poole et al., 1988, see Figure 18). The middle ear structures of the African elephant are also extremely large compared with other mammals (Soltis, 2010). The area of the eardrum or tympanic membrane is about 9 cm² (1.4 in.²) compared to that of a human of about 0.7 cm² (0.07 in.²). The mass of the malleus, incus, and stapes (278, 237, and 22.6 mg, respectively) are some 10 times that of the human equivalent. The large sizes of the elephant's auditory structures may well enhance their ability to detect frequencies down to and below 10 Hz.

The eardrum vibrates in response to a sound pressure wave and the vibrations are transferred by the three small bones known as *ossicles*. The size, shape, and attachments of the ossicle vary greatly among mammals. In general, the larger the mammal, such as the elephant, the larger the ossicles. The size of the ossicle is inversely related to frequency. The ossicles serve to both amplify the sound and transmit it to the oval window of the cochlea or inner ear. The precise relationship between the structural features of the external and middle ears and their auditory functions are not well understood in any mammal, let alone in the elephant. Sensitivity to low frequencies is, however, thought to be associated both with large tympanic membrane areas and large, compliant middle ear spaces. In *Loxodonta*, the linear dimensions of the malleus in the middle ear are about twice those of the incus. The long arm of the malleus (the manubrium) in *Loxodonta* is nearly vertical, placing it perpendicular to the horizontal plane and possibly increasing the sensitivity to low-frequency sounds (Rosowski, 1994, p. 173).

The cochlea has the characteristic spiral shape and is filled with fluid. The cochlea in *Loxodonta* has two spiral turns and a total length of the basilar membrane of 60 mm (von Békésy, 1960, pp. 506–509). von Békésy (1960) measured the resonance of the cochlea dissected from an elephant, finding it responsive down to 30 Hz. von Békésy also concluded that of the animals studied, the elephant cochlea exhibited the sharpest resonance, indicating that elephants may be well equipped to distinguish between frequencies (Long, 1994, pp. 26–27).

## SOUND LOCATION

The size of the elephant's head and ears may govern how well it can locate sounds, particularly using low frequencies at large distances. Heffner and Heffner (1982, 1984) show that an Asian elephant can localize low-frequency sounds (<1 kHz) to within an azimuth angle of 1°, which is about as well as humans can. Langbauer et al. (1991), in a playback experiment, was able to show that African elephants can locate the source of a low-frequency call at an estimated range of 2 km (1.25 mile). Sounds are localized by using interaural time differences (ITDs) and interaural level differences (Palmer, 2004). Low frequencies work best in generating a time delay between the ears, whereas highs and lows (peaks/troughs) in the sound record work best at high frequencies. The ITD for humans is on the order of 0.7 s but is unknown and probably much larger for an elephant based both on the size of the head and pinnae as well as the path traveled by the sound wave between the eardrums. The path traveled by a wave front between the eardrums is larger than the skull perimeter for low frequencies but equal to the skull perimeter for high frequencies (Kuhn, 1977, 1987). For humans, the effective acoustic circumference for low frequencies is 150% that of high frequencies (Brown, 1994, pp. 64–69). The size and position of an elephant's ears and the presence of the trunk and forehead structure would suggest that this expansion in perimeter distance for low-frequency signals might be considerably greater for elephants (Kuhn, 1977).

Whether and how animals determine range using infrasound is poorly understood. Elephant behavior with respect to low-frequency calls by familiar and unfamiliar elephants indicates an acoustic perception of range (Moss, 1988; Payne, 1998; Poole, 1996). McComb et al. (2003) used low-frequency playback calls, which provide evidence of a sense of distance between the listener and the sender of 1–2.5 km (0.6–1.5 mile). From personal observations in the field (Etosha National Park, in 1999), the time from initial reaction of elephants at a waterhole to the arrival (10 min) of a new group of elephants and the pace of the incoming group (8.0–9.6 kmph) suggest that the range at which their presence is first detected is between 1.5 and 2.5 km (0.9 and 1.5 mile). Distances in excess of about 100 m (328 ft) eliminate sight (but not smell) from the potential cues being used in determining these ranges.

However, it is important to distinguish between analog and digital characteristics of calls. Information contained within the call in analog form depends on the structure of the call. This structure, even in long-range, low-frequency calls, is attenuated more rapidly than the call itself. When information is transmitted in digital form, best visualized as Morse code, then the loss of signal (and meaning) is greatly reduced compared to the analog form and one animal can hear and interpret the call of another at much greater distances (see discussion below on estrous calls).

Only candidate theories exist that might explain how and how well elephants determine the distance of the sound source from their location: there is clearly a loss of higher frequencies with distance such that ultimately only low frequencies remain. This change in the frequency spectrum may be used by elephants to estimate range.

For hearing processes to be interpreted, we also need to know the sensitivity of the elephant's hearing at different frequencies. This quantity too is poorly known. The only existing study (Heffner and Heffner, 1982, 1984) suggests that the threshold of hearing at 17 Hz is 50 dB SPL for an Asian elephant.

The distance or maximum range over which one elephant can hear the call of another is important for a number of reasons that influence the survival of the species. In the matrilineal society of an elephant, where mature males are not in the herd and may be many kilometers away, a means to reach these males is essential. This need is heightened by the fact that the female may only come into heat for a short period every 4–5 years. Although at this point the estrous cycle may last for 4 months, the estrous periods are short (2–6 days) (Leong et al., 2003). Only when offspring are not successfully reared do females cycle over shorter intervals. Not only must the female be able to let the male know that she is ready to mate, but she must pass this message on to a number of males such that genetic selectivity can be exercised (increasing the probability that the fittest and strongest male is selected for mating). The need then for calls of females in estrous to be heard by males over a long distance and thus over a large area may well be an added reason why elephants have retained the ability to generate and hear low-frequency calls.

## THE ROLE OF THE ATMOSPHERE

For this to happen, the atmospheric conditions that were pervasive in the forests in which elephants evolved must be replicated over the dry subtropical savannas of current-day elephants. Carbon 13 data show that forests (C-3 vegetation) contracted progressively around 10 million years ago (MYA) being replaced by grasslands (C-4 vegetation). Early elephants changed fairly rapidly from feeding on C-3 vegetation to feeding on C-4 vegetation about 8 MYA. This progression from forest to grassland, with some woodland still present, resulted in mixed feeding by elephants, which changed little over their subsequent history (Lister, 2013).

As outlined in the Introduction, the dry and often cloudless atmosphere of the habitat of *Loxodonta africana* and *Elphas maximus*, the African and Indian subtropical savannas, go through a pronounced cycle of daytime heating and nighttime cooling at the surface. When translated to heat being transferred from the earth's surface to the air immediately above the surface, the heat is lost from the surface to the atmosphere before sunset until more than an hour after sunrise. This results in a strong capping nocturnal inversion forming a channel or duct in which low- (as well as other) frequency sound can be propagated. This layering of the atmosphere (cold air at the surface about 100 m (328 ft) thick and warm air above it), decouples the stronger winds higher up in the atmosphere from the surface like a layer of oil sliding over water. Calm or low winds occur at the surface during the early part of the night, gradually increasing later in the night. This increase in wind speeds as the night progresses is due to cold air drainage over sloping terrain. By dawn the cold air has pooled and the winds at the surface once again subside (Garstang and Fitzjarrald, 1999; Garstang et al., 2005).

Langbauer et al. (1991), in his playback experiment in Namibia, used a sound pressure level (SPL) at 1 m (3 ft) of 112 dB and a threshold of hearing of 46 dB, yielding a range of 1 km (0.6 mile). He then argued that because the sound production of the loudspeaker was limited to half power, the actual range of hearing was double the calculated value (i.e., 2 km (1.2 mile)).

If, however, the atmosphere is stratified and the sound signal ducted under an inversion, spherical spreading does not occur and the sound can travel much further before dropping below the threshold of hearing. Propagation of sound in a stratified atmosphere can be determined by approximating a solution to a complicated equation called the Helmholtz acoustic wave equation.

---

### The Propagation of High- and Low-Frequency Sound

The pervasively cold floor of the forest compared to the warmer canopy results in an inversion of temperature with denser air at and near the floor (lower temperatures) and less dense air at and near the canopy (higher temperatures). Sound in a fluid (air) travels faster at high temperatures and slower at low temperatures. The result in the forest is to channel or duct the sound within the canopy.

High-frequency sounds (typically above 1000 Hz or cycles per second) behave like a beam or ray of light. While such a beam of sound may propagate over long distances in an unobstructed environment, the short wavelengths (a few centimeters or less) of high-frequency sound in a forest will be intercepted by the vegetation and will quickly dissipate.

Low-frequency sound (below 20 Hz) emanates from its source not as a beam but as a uniformly propagating surface of a sphere. This too diminishes as a function of the distance from the source. However, it now decreases as the square of the distance from the source such that it loses 6 dB for each doubling of the radius of the sphere. If, however, the low-frequency sound is contained within a duct (the inversion of the forest canopy), it will propagate down the duct, decreasing as a function of the linear distance from the source and not as the square of the distance from the source, traveling now like the high-frequency sound as a ray of light. But with a huge difference: The low-frequency infrasound has a wavelength of greater than 17.5 m or 40 ft compared to high-frequency sound with a wavelength of a few centimeters or less than an inch. The long wavelengths of infrasound simply go "around" the trees and vegetation whereas short wavelengths of high-frequency sound crash into the trees and vegetation and are dissipated.

As elephants evolved in the forest they took advantage of the "inversion or ducting" conditions, developing the capability of generating and detecting low frequency or infrasound. They subsequently emerged from the forests onto the savannas with this infrasonic capability. The amazing fact is that these hot, dry savannas, which during the day destroy almost all sound propagation, flip sides before sunset until after dawn to create a pronounced duct or inversion at the surface ideally suited to the propagation of infrasound.

This equation has no solution and must be approximated using computers to successively arrive at an acceptable estimate of a solution. Because helicopter rotors generate infrasound, they can be detected when flying below radar (terrain) surveillance height, which simply means flying below the height of the surrounding topography. The Department of Defense and the National Acoustics Laboratory in Mississippi spent a large amount of time and money developing the Fast Field Program (FFP) (Franke and Swanson, 1989; Lee et al., 1986; Raspet et al., 1985) that would yield a timely and acceptable result. These researchers, in collaboration with scientists at the University of Virginia (Garstang et al., 1995; Larom et al., 1997), then applied the FFP to elephant communication in the presence of these nocturnal inversions.

## RANGE OF ELEPHANT CALLS

In the presence of a low-level (surface) inversion, one elephant can hear a loud, low-frequency call of another elephant 10 km (6 mile) away. In the middle of the day, however, this distance may be reduced to less than 1 km (0.6 mile). These calculations were made for calm wind conditions. Calm or low-surface wind speeds typically occur under strong inversion conditions. Strong winds occur around the middle of the day when heat-driven turbulent mixing is at a maximum (Figure 9.5). The effect of wind on the propagation and hearing of sound is complex. Turbulent motions in the atmosphere typically form vortices and rolls in the atmosphere that are highly destructive to sound of almost all frequencies. Standing on a bridge over a fast-flowing river, the eddies and whirlpools visible

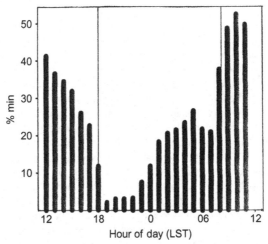

**FIGURE 9.5** Average frequency (%) of wind speeds higher than $4\,m\,s^{-1}$ (2.1 mph) (thick black lines) in each hour for the 17 days in eastern Etosha National Park, Namibia. The thin vertical lines delineate net daytime heating (08:00–18:00) from net nighttime cooling (18:00–08:00) at the surface. *After Garstang et al. (2005).*

at the water's surface, gives one a visual picture of what occurs in the atmosphere. Turbulence is certainly the most destructive atmospheric effect on the range of low-frequency sounds used by elephants. Turbulence near the surface is closely related to wind speed, with higher turbulence occurring in higher winds. Since friction constrains wind to be zero, at some point very close the surface, wind speeds increase with height producing layers of faster-moving air over layers of slower-moving air. This change in wind speed with height produces a shearing motion of one layer of air over another, enhancing mixing and turbulence. Wind-driven refraction of sound shortens upwind and lengthens downwind communication. Finally, wind creates noise and in particular flow noise across the elephant's ears. Hunters and game-watchers know that animals tend to seek shelter and lie low on windy days. This may be mainly due to the fact that they have difficulty hearing and thus detecting danger (predators) on windy days.

The net result of the relationship between thermal stratification, wind, and the propagation and detection of sound is that one can conclude that optimum communication conditions occur when both thermal stratification and low winds coincide (Figure 9.6).

The area over which an estrous call being made by a female elephant can be detected by a male governs the number of males reached by that call. Under poor daytime conditions, this area may be less than $3\,km^2$ ($1.9\,mile^2$); under the best early evening or morning conditions, the area will be more than $300\,km^2$ ($118\,mile^2$). In the first instance of less than $3\,km^2$ ($1.2\,mile^2$), no males may be found. In the second instance, a number of males will be able to hear the call.

Researchers have been unable to find any distinctive structure to the loud estrous calls made by elephants. Payne (1998), however, found that females in estrous called more frequently than at other times and that males may recognize this pattern of calling rather than deciphering the content of the call. The mating pandemonium emitted by both the female in estrous and her family, as described earlier in this chapter, would, however, represent a distinctive signal to distant males. McComb et al. (2003) found that calls around $100\,Hz$ contained a complex pattern of formant frequencies spanning several harmonics that identified the caller. They concluded that the content of these calls was critical to social recognition and identification of the caller. The playback experiments were conducted between 07:00 and 13:00 h but were not registered against sunrise or details of the near-surface temperature and wind fields. Wind speeds were reported to be low (7 mph) but no information was given on how and at what height these winds were measured, nor whether gusts exceeded the 7 mph threshold. McComb et al. concluded that content was as important as the range of the caller and that social recognition was thus probably limited to less than 2.5 km (1.5 mile). These conclusions, while valid for the times and place that the experiment was conducted, cannot be generalized over 24 h and are likely to be substantially different at other times within the 24 h.

More importantly, however, is that if, as both Payne (1998) and Soltis (2010) have suggested, females before or during estrous call at a characteristic rate

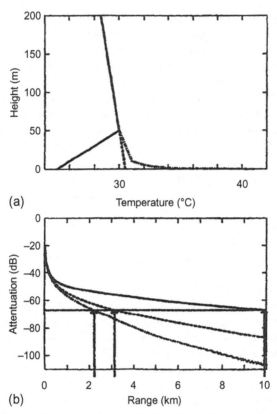

**FIGURE 9.6** Idealized temperature structure for three times of day: (a) dotted line: midday, dashed line: transition between daytime heating and nighttime cooling, solid line: nocturnal inversion, and (b) the resulting range or distance for each temperature profile that one elephant can hear another elephant's loud (117 dB) low-frequency (15 Hz) call with a threshold of hearing of 49 dB. *Based on Larom et al. (1997).*

(i.e., make a digital call), these calls are independent of loss of content and can be heard and interpreted at much greater ranges. Using a calling pattern that transmits crucial information over the greatest distance, independent of call content, may have found its origins in the dense forests in which elephants evolved.

## TIMES AND FREQUENCY OF CALLING

It has been suggested above that elephants, particularly mature females, may produce low-frequency calls on an almost continuous basis. Only Langbauer and Payne (Langbauer et al., 1991) have attempted to record calls of elephants on a continuous basis (Figure 9.7). They collared 14 female elephants in the Sengwa Reserve in Zimbabwe carrying sound recorders that would record only the loudest low-frequency calls. Calls below a given sound-pressure level were

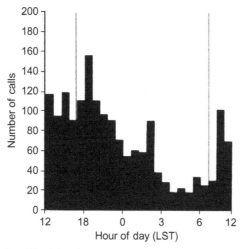

**FIGURE 9.7** Number of loud, low-frequency calls made in each hour recorded from 14 collared adult elephants (*L. africana*) in the Sengwa Reserve in Zimbabwe. The thin vertical lines delineate the daytime heating and nighttime cooling as in Figure 9.5. *Langbauer and Payne (personal communication, 2000); Garstang et al. (2005).*

not recorded. Thus, a full spectrum of calls is not available, yet the record of loud, low-frequency calls is known.

These collared elephants show a clear maximum in calling in the early evening at a time when nocturnal cooling, inversion formation, and low wind speed coincide. As the night progresses, calling rates decline starting once again in the second hour after surface heating begins. Loud, low-frequency calling during the day continues at an average rate per elephant of about eight calls per hour, or one loud call every 8 min.

Garstang and his colleagues (2005) placed eight microphones around a remote waterhole (Mushara) in the eastern end of the Etosha National Park in Namibia. A continuous record of all the calls recorded was analyzed for 8 consecutive days in September 1999 (Figure 9.8). This record differs in a significant way from that

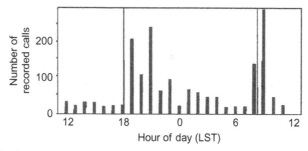

**FIGURE 9.8** The number of elephant low-frequency calls (<100 Hz) recorded over 8 consecutive days (13–20 September 1999) in each hour in eastern Etosha National Park. Thin vertical lines as for Figure 9.7. *Based on Garstang et al. (2005).*

of the collared elephants. In that record, it is the number of loud, low-frequency calls made by 14 individual elephants. In the Mushara record, it is the number of calls heard or detected. Since calling activity is probably contagious, such that elephants call more often when they hear more calls, the observed distribution of calls at the Namibian site will be a product of the proximity of elephants to the recording site (the waterhole). The number of elephants calling, the calling behavior of these elephants, and the number of calls that can be detected by the microphones are all a function of the prevailing atmospheric conditions. Under optimum acoustic conditions (strong low-level inversion with no wind), the microphones at the waterhole may detect calls of animals in a surrounding area of $300 \text{km}^2$ ($118 \text{mile}^2$). Under the worst acoustic conditions (strong surface heating and surface winds), this area will have shrunk to a less than $3 \text{km}^2$ ($1.2 \text{mile}^2$).

The distribution of calls recorded over the 24 h at Mushara is thus dramatic: 42% of the 1400 calls recorded occur in the 3 h following sunset (19, 20, 21 h, local standard time) and 29% are recorded in the 2 h following sunrise (08, 09), for a total of 71% of all of the calls recorded in 24 h. Both of these peaks in calls recorded occur at times when atmospheric conditions are at or near optimum for the transmission of low-frequency sound. Of the remaining 30% of all detected calls, 25% occurred at night, leaving only 5% for the daytime hours. While the observed distribution of calls recorded reflect the presence and absence of elephants at a watering site, the calls are recorded most often when the call travels the furthest.

These observations suggest that elephants also make more calls when they hear more calls. The number of calls recorded by our microphones during times of optimum atmospheric acoustic conditions suggest that the calling rate is influenced by the number of calls heard. Such a hypothesis would need careful identification of both caller and receiver, perhaps to the extent that the precise timing and location are known.

Payne (personal communication, 1995) was able to track the movements of herds within communication ranges of each other. She noted, in particular, that adjacent herds while approaching each other never cross paths. To do so would be energetically costly. A given adult elephant consumes between 150 and 200 kg of vegetation in each 24 h. At the end of the dry season a herd of elephants consisting of 10–20 adults plus young would largely denude the area over which they are feeding of most of the edible vegetation. A second herd coming into this area would find little to eat. Payne's observations suggest that the signals of each herd are used by the other to modify their feeding pattern. It is quite possible, given earlier discussion (Chapter 5) of the ability of elephants to recognize other elephants based entirely on call recognition, that elephants in one herd are aware of the number and composition of an adjacent herd.

Researchers in the field elect to work in daylight. Far fewer observations are made at night yet it is clearly imperative that it is the cycles of the natural biological and physical world that should be considered rather than entrenched human behavior. Increasingly, we are able to measure variables remotely and continuously. Our defective field observing systems will progressively improve, although the reluctance of the natural scientist to recognize the role of the physical world and document its behavior concurrent with that of the living world is still in need of substantial improvement.

## ABIOTIC SOUNDS

Elephants with their exceptional sense of hearing detect signals from the biotic as well as the abiotic world. Both O'Connell-Rodwell et al. (2001, 2004) and Hägstrum (2000) have reported that animals may detect and use seismic signals. Garstang (2009) has suggested that in the wake of the Sumatran earthquake on 26 December 2004, elephants in both Sri Lanka and Thailand, 1000 km (620 mile) away, were able to detect the sound in the atmosphere generated by the tsunami wave crashing on the shores of Sumatra.

Sound in the atmosphere travels slightly faster at sea level than the speed of the tsunami (1200 km h$^{-1}$ (750 mile h$^{-1}$) vs. 700 km h$^{-1}$ (440 mile h$^{-1}$)). At 1000 km (620 mile) from Sumatra the sound wave would arrive a little less than 40 min before the tsunami struck. Anecdotal evidence indicates that elephants in both Sri Lanka and Thailand responded 20–60 min prior to the arrival of the tsunami. This 40 min advance notice of the sound wave falls within this 20–60 min time window. Elephants on the beach in Thailand had just returned from giving tourists rides. They were chained to stakes driven into the ground just off the beach. These elephants were reported to have screamed, broken the restraining chains, pulled the stakes out of the ground, and run to high ground all within the above time frame of 20–60 min.

Two other potential cues could have alerted these elephants on the beach to the threat of a tsunami. Water along the shoreline and most noticeably on a gently shoaling beach withdraws some 20 min before the arrival of the tsunami. Such withdrawal is in response to the trough ahead of the wave itself and would create both an unusual sound as well as an unusual smell. Both signals could have been detected by the nearby elephants. Whether and how the elephants acquired memory of such precursor events is unknown. It is possible that precursor signals in the earth's crust (S-waves, Love, and Rayleigh waves), which would all have arrived within 15 min of the earthquake and more than an hour before the tsunami, could have alerted, but not panicked the elephants. The combination of sound and smell signals, however, may well have triggered a response (Garstang, 2009).

Kelley and Garstang (2013) have shown that infrasound produced by thunderstorms generates sound waves with pressure levels that elephants can detect at distances as great as 150 km (93 mile) from the storms.

Lindeque and Lindeque (1991) have suggested that elephant herds in eastern Etosha National Park in Namibia head toward the Caprivi Strip 2–3 weeks before any of the other herd animals such as wildebeest and zebra begin to move.

It is now possible to speculate that these movements are initiated by the audio detection of remote thunderstorms heralding the end of the dry season.

Research at the University of Virginia, together with work being done at the universities of Utah, Texas A&M, and Cornell, is exploring this relationship between elephant movements and the occurrence of rainfall. Results show that a distinct shift in movement of the herd occurs when rain begins to fall after a prolonged (month's) dry season at a location hundreds of kilometers from these elephants (Garstang et al., 2014; Kelley and Garstang, 2013).

Garstang and his colleagues (2014) and Kelley and Garstang (2013) suggest that elephants detect the low-frequency sounds generated by these distant rainstorms, know that these signals mean that the wet season rains have started, and change their movement behavior in response to these signals. The elephants studied by Garstang and his colleagues in the far western Kunene region of Namibia did not exhibit major changes in movement such as migrations out of the area toward the rains, but rather showed changes in direction and distances traveled in daily movements.

It is entirely possible that not only are elephants aware of the distant rainstorms but that they are also aware of the relationship between rainfall, river catchments, and runoff in the ephemeral rivers of northwestern Namibia and the greening of the vegetation.

Garstang et al. (2014) further found that although the elephants change their movement patterns in apparent response to distant rainstorms, these changes are not consistently reflected in all of the elephant herds that were tracked nor in fact responded to by all herds with members carrying GPS collars. This absence of a consistent and uniform response emphasizes the difficulties faced when attempting to understand animal behavior, once again reflecting that absence of evidence is not evidence of absence.

Cyril Christo and Marie Wilkinson, in their book *Walking Thunder* (Christo and Wilkinson, 2009), relate a Turkana legend from northern Kenya in which the sighting of an elephant is a sign that rain is imminent. They also found that further south in Kenya, the Samburu people believe that elephants know when rain is coming. The Samburu say that the sudden reappearance of elephants, after months of no rain, signals the coming of the rains. Christo and Wilkinson also note that in India the elephant was believed to bring the monsoon rains and considered the elephant to be allied with cumulus clouds. The insight displayed by people living close to animals reflects the depth, if not the explanation, of their observations. We should not discard the observations of these various peoples because we perceive the explanation offered to be lacking or inadequate.

A lone female African elephant, possibly a forest elephant (*Loxodonta cyclotes*), who came via Brussels, Belgium, had wound up in the Lahore Zoo in Pakistan (R. Garstang, personal communication, 2002). It is likely that this elephant had never had a companion and never vocalized unless a 747 Boeing aircraft took off at the Lahore airport some 15–20 km (9–12 mile) away.

She then responded with low-frequency rumbles, almost certainly triggered by the wake vortices of the 747 engines, which generate considerable infrasound.

Poole and her colleagues (2006) report that a 10-year-old female African savanna elephant living some 3 km (1.9 mile) from a Kenyan highway imitates the low-frequency engine sounds made by heavy trucks. The sounds she makes statistically match the engine sounds and are different from normal elephant calls. Poole also reports on a 23-year-old African elephant living in captivity with two female Asian elephants who has learned to chirp like the Asian elephants.

These findings not only demonstrate that elephants are capable of vocal learning, imitating signals not typical of their species, but suggests that their vocal learning capabilities reflect selective evolutionary pressure that affects their social relationships.

Elephants that have been exposed to culling where helicopters and firearms have been used are fully aware of the meaning of the sounds generated by these sources. In the Kruger National Park, in particular, culling operations were conducted near sunset in order to take advantage of cool conditions. We now know that this is the very worst time of day for such a traumatic event to take place. Not only would elephant herds within a radius of at least 10 km (6 mile) hear the sounds of distress and panic of their fellow species, but they would hear the infrasound from helicopter blades and gunfire over distances of perhaps greater than 100 km (60 mile). They would thus be all but witness to this traumatic scene. Payne and Martin (personal communication, 1995) report that in a culling operation in the Sengwa National Park in Zimbabwe, elephants hearing the sounds of the operation fled 145 km (90 mile) to the borders of the park.

That elephants can detect the seismic field of the earth and use this signal to navigate remains a possibility that awaits further research.

The huge ears of the African savanna elephant hearing system also serve the function of dissipating heat in environments where daytime temperatures can exceed 40 °C (104 °F). Heat loss through the blood vessels near the surface of the ears is increased by flapping of the ears. For animals weighing as much as 7 metric tons, however, this may not be enough. Recent work has shown that other areas of the elephant's body contain dense networks of blood vessels to which the flow of blood may be selectively controlled and used as locations for heat dissipation. Weissenbock and her team at the Vienna Zoo have shown surface temperature differences on the elephant's body to differ by as much as 20 °C (68 °F). They suggest that elephants may be able to consciously open and close these thermal windows (Weissenböck et al., 2010, 2012). Other work being conducted at Princeton University by Myhrvold suggests that the sparse covering of bristly hairs found on an elephant might act as "tiny heat fins." A computer model developed by Myhrvold to test this idea suggests that hairs can boost the heat loss from an elephant's body by 20% (Myhrvold et al., 2012).

Elephants may also take advantage of the extremely low temperatures experienced on the open dry tropical savannas at night. As decribed earlier in the Introduction, low humidity and cloudless skies in savanna elephant habitat result in huge and rapid radiative losses of heat from the earth's surface to space. Surface temperatures may drop from daytime highs of above 40 °C (104 °F) to less than 5 °C (41 °F) beginning before sunset. Elephants may allow their body temperature to drop abnormally low under these cold nighttime conditions in order to cope with the extremely high daytime temperatures.

The demands made on the brain to process the amount and complexity of auditory and other signals is considerable. Not only are a multitude of signals taken in by the neural network, but these signals must be processed simultaneously in a variety of ways. Memory sources need to be combined to identify acoustic sources with complex discriminatory functions that must determine the identity and location of these sources. These operations are further extended to incorporate rudimentary language.

Pijanowski and colleagues (2011), as well as previous researchers (Krauss, 1987; Southworth, 1969), have suggested that sound is fundamental to the ecological landscape and as such must be incorporated as an essential part of the environment on that scale. The value of this vision is that sound in the animal world is much more than communication between and among species. Rather, sound from nonbiological sources may be actively used in decision making (movements), area occupied (home range), navigation (migration, resources), and avoidance of danger (fire, flood) (Pijanowski et al., 2011).

# Chapter 10

# Language

We have suggested that elephants communicate with each other, perhaps using a far more complex system than we do, incorporating sound, smell, sight, taste, and touch. The question is whether any of this exchange translates into what could be called language with a vocabulary and structure. *Webster's Collegiate Dictionary*, for example, defines language as "any means, vocal or other, of expressing or communicating feeling or thought." There is little doubt that elephant communication meets such a definition.

In elephants, at least 10 different vocalization classes have been identified (Clemins et al., 2005). Using speech processing techniques and seven captive adult African elephants, Clemins and colleagues (2005) were able to identify each of these vocalization types with an accuracy of between 80% and 94% depending on the quality of the dataset used. The elephant emitting the sounds could be recognized nearly 9 out of every 10 times.

Poole (2011) recognizes 13 call types divided into three groups based on the source of the sound (Table 10.1). Seven rumble calls are classified as laryngeal calls, three as trunk calls, and three as initiated or idiosyncratic calls. The calls may have a range of meanings, which Poole calls "gradiness," as made by the caller but may have a discrete meaning as interpreted by the receiver. Poole recognizes 35 contextual calls, many of which have a very discrete meaning such as the "let's go" rumble. She divides these 35 calls into eight contextual classes.

Poole (2011) indicates that there might be a range of meanings to the call emitted but a discrete interpretation is made by the receiver. Defensive calls may result in immediate reaction by all members of the herd gathering in a tightly knit group or simply responding to where two herds are in contact with each other. Payne (1998), for example, reports that GPS-collared groups feeding in the same area and tracked over a number of weeks never crossed each other's paths. Such behavior avoids expenditure of energy in going into an area already depleted in resources.

Communication between elephants over considerable distances indicates the ability to locate the caller with considerable accuracy. Playback experiments of female estrous calls (Langbauer et al., 1991) demonstrate that males can locate the source of the recorded call 2 km (1.2 mile) away. Palmer (2004) points out that while elephants are capable of using interaural time delays to determine the location of calls, this process requires considerable neural processing. A network

Elephant Sense and Sensibility. http://dx.doi.org/10.1016/B978-0-12-802217-7.00010-7

**TABLE 10.1** Sources of Elephant Sounds

| **Laryngeal Calls** | |
| --- | --- |
| 1 | Rumble |
| 2 | Rev |
| 3 | Roar/roar rumble |
| | Tonal roar |
| | Noisy roar |
| | Mixed roar |
| | Pulsated roar |
| 4 | Cry/cry rumble |
| 5 | Bark/bark rumble |
| 6 | Grunt |
| 7 | Husky cry |
| **Trunk Calls** | |
| 8 | Trumpet |
| | Pulsated trumpet |
| 9 | Nasal trumpet |
| 10 | Snort |
| **Imitated and Idiosyncratic** | |
| 11 | Truck-like |
| 12 | Croak |
| 13 | Squelch |
| **Context Calls** | |
| Multiple subcall types related to: | |
| Group defense | |
| Food | |
| Sexual | |
| Mother–offspring | |
| Conflict | |
| Social integration | |
| Logistical calls | |
| Play | |

*Modified after Poole, 2011.*

of neurons must act in the elephant's brain to detect the moment when the signal is received simultaneously in each ear, identifying the direction probably within $1°$ of azimuth of the location of the source. Depending on the frequency of the sound being processed by the brain, it is possible that further processing involves phase changes in the signal to achieve the necessary discretional resolution.

The work referred to earlier by McComb and colleagues (2000) (see Chapter 5) described how a single elephant, often the matriarch, can recognize and identify 100 other elephants in the wider population. This ability calls upon considerable neural capacity, including the ability to store and recall a large amount of information.

Seyfarth and Cheney (2003b) as well as other psychologists maintain that no animal other than humans calls to inform another animal or that the caller intends to provide information to the receiver. Contrary to these assertions, current work shows that contact calls between elephants allow the caller and receiver to identify each other, resulting in an observed response such as each converging on the other. Similarly, the "let's go" rumble initiates a change in the behavior of the receivers (i.e., an interpretation of the content of the call). Equally, accumulating evidence in animal communication suggests that exchanges between callers and receivers go well beyond the receiver acting as passive eavesdropper.

An example of a sequence of exchanges between a mother and her calf in the Addo Elephant Park is described by Watson in his book *Elephantoms* (2003). The mother and calf had been separated by a fence that was being repaired. As related to Watson by a ranger who was at the scene, the calf responded to inaudible instructions from the mother to move into cover (deep shade) about 20 yards on the other side of the fence. The calf did this and stood waiting motionless and almost invisible in the deep shade. The mother then rushed through the gap in the fence and onto the road, confronting the fence repair crew in an oncoming truck. In the turmoil of the charge and braking of the truck, both of which generated clouds of dust, the calf slipped through the gap in the fence into the dense bush of the park.

This event leaves us with at least two issues to ponder: Did the mother figure out the detailed strategy that led to the calf's escape? Was the mother able to communicate the essential details of this plan to her calf? The sequence of events and deliberate actions and responses of each animal make chance a poor candidate solution.

Masson and McCarthy (*When Elephants Weep*, 1995, pp. 64–65) relate the story of MaShwe, a work elephant, and her 3-month-old calf caught in the rising flood waters of the Upper Taungdivin River in Burma (Myanmar). MaShwe was able to keep her footing but the calf was in danger of being swept away. Despite the mother's efforts, the calf was swept downstream. MaShwe plunged after her and managed to pin the calf against the bank with her head. She then lifted the calf in her trunk, reared up on her hind legs, and placed her on a rocky ledge 5 ft above the flood water. No sooner had MaShwe managed this when she fell back and was swept away by the torrent.

Some half hour later as the elephant handlers, who had witnessed the entire drama, were unsuccessfully trying to reach the calf, they heard a "defiant roar" from MaShwe as she appeared on the opposite bank. As soon as she saw her calf her calls turned to low rumbles, calming the calf.

She was not able, or was consciously unwilling, to cross the river until morning when the flood had begun to subside. MaShwe then crossed the river and rescued the calf from the ledge.

Under these extreme conditions of stress, the mother was able to take coherent action, respond to unexpected crises, return to the correct location, warn off the humans who were getting involved, change her mode of communication to calm the calf, keep the calf calm despite its precarious situation throughout the whole night, and then under less dangerous conditions rescue the calf.

The entire sequence of events calls upon considerable rational behavior together with purposeful communication, signaling clear intent on the mother's part and an expected response by the calf. Describing the event in terms of instinctual parental/offspring response would seem to be a poor alternative explanation.

We, as humans, are justifiably impressed by the incredible complexity of our own language and our ability to use it. Somewhat blinded by this fact, we have failed to recognize the ability and complexity of language and communication used by other species. Not knowing what is being said and how others are using their language severely limits our ability to assess the capability of others and in so doing we run the risk of ignoring them.

It is informative to reflect on what is known about the evolution of language in humans. Most linguists anchor the origin of language to the emergence of symbolic thought as it appeared for the first time in engravings and artifacts crafted by early hominids. Geometric patterns engraved on an artifact found in Blombos Cave, in the Cape Province of South Africa, is the earliest example of such symbolic art. The Blombos artifact dates to only 70,000–80,000 years ago (YA) (Figure 10.1) (McCarthy and Rubidge, 2005). At about 100,000 YA, distinctive evolution in the vocal tract of early hominids took place. The FOX P2 gene in the Hadzabe people of northern Tanzania appeared and aided in speech formation and motor coordination. A "click" language may have been developed and remains in evidence in the speech of the Hadzabe and San (South Africa, Botswana, and Namibia) people. The click language is based on five different clicks and consists of only about 140 distinct sounds. Most words are only a single syllable and meaning is derived from the order of clicks and not inflection. The characteristics of this early language in hominids are not that far removed from the current status of language in elephants.

There are numerous reports of elephants drawing and painting, including those in Thailand where the work is sold to tourists. Despite the possibility that the Thai elephants may have received some form of instruction and certainly receive rewards for their paintings, the paintings are nevertheless remarkable. Mark Freeman's demonstration seen in January 2014 of elephants painting in

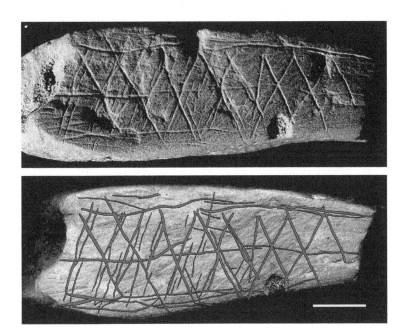

**FIGURE 10.1**   An engraving in red ochre on a clay tablet from Blombos cave on the southern Cape coast. The specimen is 77,000 years old and provides the earliest evidence of cognitive abilities central to modern human behavior. (Top) Clay tablet as it appears to the camera; (bottom) red ochre scratches. *From McCarthy and Rubidge, 2005. Permission courtesy of Professor Christopher Henshilwood, University of Bergen, Norway.*

Thailand must give the neuropsychologists considerable pause for thought. The images painted are unquestionably those of elephants.

The execution in which successive brushstrokes made with apparent confidence, combined to create a coherent composition, raises serious questions as to the origin and conception of the ultimate product. Does the elephant have such a completed image formulated in her mind? If so, such an achievement calls upon some of the higher forms of neural processes. Observing the actual creation of the image by the elephant strongly suggests that the equivalent of a mental picture preexists. The initial stroke of the portrait recorded by Freeman was not only elegant in composition and execution, but was clearly the foundation for the ultimate creation.

Siri, a young Indian elephant, drew spontaneously in the dust of the floor of her enclosure. She manipulated a pebble under her foot to scratch out designs on the floor. She would then trace her design with the tip of her trunk, perhaps detecting the smell of the exposed concrete or the residual flakes of the pebble. Siri can draw with a pencil held in her trunk within the bounds of a $9 \times 12$ in pad of paper held by her handler, Gucwa, on his lap. The fluid, flowing drawings seen by artists not knowing who drew them were regarded as equal to and often superior than those done by human artists. No rewards or prompting were given to Siri either for her drawings on the concrete floor or on the pad of paper (Masson and McCarthy, 1995, pp. 205–207).

As recent as the Blombos artifact is in evolutionary terms, it took nearly all of the 80,000 years for humans to move from a spoken to a written language. The Rosetta Stone dates from 196 B.C. with the possibility that Early Stone Age art was in fact a form of written language. The pace of human development of language since the appearance of the first written words is phenomenal. The power now being unleashed even more rapidly by the expanding global capacity to communicate can only be guessed at. This complexity of the human capacity to communicate makes it even more difficult to recognize and interpret the extraordinary abilities of animals to perform feats of communication that we as humans might find difficult to achieve or even fully comprehend.

# Chapter 11

# Intelligence

Earlier we defined intelligence as the "capacity to meet new and unforeseen situations by rapid and effective adjustment of behavior." While this definition may serve to characterize intelligence, the challenge is how do you apply this definition to an elephant and, as important, how do you evaluate the result? It is difficult to determine what constitutes intelligence or intelligent behavior without using human behavior as the standard (Byrne, 2006). The ability of elephants to recognize and remember some 100 other individuals is not only a mental feat that may not be matched by all humans but that recognition by elephants is not the product of sight but of sound and enhanced by other senses such as smell and touch. It is very likely that if humans were asked to demonstrate their ability to recognize by sight more than 100 individuals whom they have identified by sound, the result may fall far short of what elephants can do.

We are faced with the additional problem of deciding the level of intelligence involved in learning and the role that learning plays in any mental exercise. The definition we quoted above attempts to eliminate the role of learning by specifying that intelligence is the "capacity to meet new and unforeseen situations" and to solve the potential problem by "rapid and effective adjustment of behavior." But if, as we have argued earlier, the brain has evolved over evolutionary time to promote survival by adapting to the challenges of the environment and to retain in its unconscious those strategies that have worked, then how do we know that the problem that is designed to test the elephant's intelligence is, first, "new" and, second, of any relevance to the elephant? If instead intelligence is the product of an infinite number of successful experiments, determined over evolutionary time as a result of trial and error, then there is no moment of discovery or capacity to meet a "new situation." If most of what we do and what an elephant does is embedded in the unconscious and called upon by the conscious to act in any given situation, then it is not possible to determine either what is "new and unforeseen" or what meets these demands for an elephant. While knowledge is stored in the unconscious and is manifest in behavior such as using the trunk to feed, that knowledge was initially consciously learned before being stored in the unconscious. It would seem that careful observation of elephant behavior under natural conditions would more likely yield evidence of intelligent behavior.

Elephant Sense and Sensibility. http://dx.doi.org/10.1016/B978-0-12-802217-7.00011-9

Elephants can locate the source of sound with considerable accuracy. As discussed in Chapter 9, one elephant can locate another by its call to a 1° accuracy in azimuth at a distance of 2 km or greater. When hearing the approach of other elephants, again at distances of 2 km or greater, elephants will frequently stand facing the direction of the oncoming elephants, with their heads raised and the ears widespread. At other times, often with the approach of a predator or even a much smaller animal such as a tortoise, an elephant will straighten and point with its trunk. Prior to flight in a "fight-or-flight" situation, the elephant that decides to flee points with its trunk the direction it will take.

Smet and Byrne (2013) have recently demonstrated that African elephants, trained to take tourists for rides, were able to follow humans pointing at food in a container with an accuracy of 68%. One-year-old children only do slightly better at an average of 73% of the time. The elephants Smet and Byrne worked with were never trained and in their previous routines were never exposed to pointing by their handlers. In fact, they performed at the above level from the beginning of the experiment and did not improve over the course of the experiment. This suggests that the elephants were making use of an inherent ability and not an acquired one.

When the experimenters only used subtle movement of the head and eyes to indicate which of the vessels contained food, the elephants failed to respond. As in the case of the mirror test (Chapter 8), it is surprising that elephants with poor eyesight and heavy reliance upon sound and smell are able to respond to visual cues. Conversely, it is not surprising that they probably failed to detect subtle head and eye movement as cues to locate the food.

Hutto (*Illumination in the Flatwoods*, 2006, p. 110) records that the young wild turkeys he guided through the flatwoods of the northern Gulf coast of Florida would respond to his pointing at an object or insect and readily know where to look for the object.

Plotnik and colleagues (2011) reworked a classic 1930s experiment used on primates to subject 12 male and female Asian elephants at the Thai Elephant Conservation Center in Lampang, Thailand, to a challenging situation that none of the elephants involved had previously encountered. A sliding table holding bowls of corn was separated from the elephants by a transverse net. The table had to be approached down two lanes, which led to the ends of ropes attached to the sliding table. However, both ropes had to be pulled toward the net for the sliding table to advance, bringing the bowls of corn within the reach of the elephants. If only one rope was pulled, it would simply slide through two rings without moving the table. The elephants quickly learned that the task had to be coordinated. They would wait up to 45 s for a partner to show up. Two elephants, Neua Un and Jo Jo, learned that it was not necessary for both of them to pull on the rope. Neua Un simply stood on her end of the rope and allowed Jo Jo to do the pulling. Furthermore, Jo Jo would not even walk up to the net unless his partner was released to join him.

This test not only involved comprehension of relatively complex mechanics but required coordination between two individuals. As constructed, the task could not be completed by a single individual. Once recognized, individuals did not attempt to complete the task on their own but waited for help. In this experiment the elephants went beyond their human task masters, finding a solution that the humans had not thought of by one elephant standing on the rope.

Holdrege (2001) recounts an incident in India that illustrates both learning as well as response to an unexpected situation. A work elephant had been taught to pull a tall pole from a truck that the mahout and elephant were following and place the pole upright in a previously prepared hole. The process proceeded down the line of holes until the elephant refused to lower the pole from midair above the next hole. The mahout got down from the elephant to find a dog sleeping in the bottom of the hole. Not until the dog was chased out of the hole would the elephant lower the pole into it. Two levels of intelligence are evident in this account. First, that the elephant is capable of learning and performing a task— the planting of the poles—and, second, that the elephant was aware of what he or she was doing. It was not acting purely by rote but perceived that with the unexpected presence of the dog at the bottom of the hole, something bad was likely to happen and that to avoid this, the elephant had to disobey commands it had been following.

Poole and Granli (2004) have examined elephants less than 7 years old at play. At this age the behavioral repertoire of elephants is still developing. For example, they have yet to master the use of their trunk, which at this early stage is seen as a rather useless appendage that is often in the way (even stepped upon). Play may even imply cognitive recognition between reality and pretense. They identified five categories of calf play that promoted motor skills, especially with the trunk, and social skills that included rules and procedures in attaining and maintaining rank and dominance. Extensive use of vocalization in play developed later communication skills. Poole and Granli noted that elephants engage in what can only be termed as absurd, even preposterous, solitary play. In this kind of play there were many instances noted of "pure expressions of joy, of fun and clowning around." Such lone performances may be examples of self-awareness in elephants. Collective play in groups, especially in water, promotes social skills and sociability. Playing in water involves a lot of body contact and close-quarter play (Figure 11.1).

Elephants frequently dig for water in dry stream beds in the dry season. Both personal experience and that of Payne (personal communication, 1995) suggest that elephants (usually the matriarch or an older adult female) seldom dig for water without finding it. Furthermore, they locate water at a depth of no more than about 1 m (3 ft). Digging in a sandy, dry riverbed, they scoop the sand out with a front foot, cupping the foot to make a scoop. They do not use their trunks in this digging operation. In many instances, water seeping into the hole will take a number of minutes to accumulate and clarify. Members of the herd will patiently wait their turn to drink, taking care not to collapse the hole,

**FIGURE 11.1**  Young elephants show an infinite variety of play from solitary to group play, involving running, mock charging of bushes and birds, great joy in water, and sometimes boisterous play that needs to be controlled by adults.

and will prevent and even assist young to drink from the hole (see Chapter 6). Elephants have been seen to cover up these artificial waterholes with vegetation, preventing other animals from getting to the hole. When not protected in this way, other animals, particularly Cape buffalo, wildebeest, and zebra, quickly collapse and destroy the hole, often before a single animal has had the opportunity to drink.

In at least one case reported by Holdrege (2001), an elephant, after digging a hole in a sandy river bed to reach water, stripped bark from a nearby tree, chewed it into a large ball, used the ball to plug the hole, and covered it with sand. Later, this elephant was observed to return to the covered hole, remove the sand and plug, and drink from the hole. The level of cognitive comprehension in this case is remarkable. Not only must the elephant recognize why the hole needs protection, presumably gained from previous experience, but he uses his tusks as tools to manufacture a device to protect the source of water that he has created and recognizes further that the protective device itself must be disguised for his efforts to succeed. Whether or not these actions are interpreted in the terms presented above cannot detract from the actual events that took place.

In a situation analogous to the account of sharing a small source of water described earlier, a BBC film crew (17 March 2011) waiting at a stagnant water hole for the annual flood to arrive in the Okavango, observed similar apparently carefully considered behavior. The shallow water, only inches deep, covered a thick layer of mud and sediment. Any disturbance would sully the water, making it unpalatable or even undrinkable. The filmmakers, like in the Wanki case (see Chapter 12), anticipated that the approaching elephants, traveling in temperatures still near 50 °C (122 °F) after probably 24 h without water, would rush into the water and destroy the precious source. Instead, the lead elephants slowed the pace, approaching quietly and stepping into the pool carefully making as little disturbance as possible; each elephant, one after another, "began to carefully sweep their trunk tips across the surface, delicately siphoning the few centimeters of clear liquid from the mud below."

As before, the behavior exhibited by a group of animals ranging in age from adults to calves suggests considerable exercise of knowledge, control, and discipline. One would presume that experienced adults or at least the matriarch of the herd would know from experience that to avail themselves of drinkable water they had to, as a group, behave in the manner observed. Under normal circumstances elephants at a water hole will typically quite violently disturb the water. In this case, however, they apparently recognized that this would not be appropriate. Instead, they adopted a completely different behavior. Such a change in behavior implies considerable neural processing including the requirement that such change in behavior had to be communicated to the entire herd.

Elephants are fully aware of an electric fence and in Etosha (Conrad Brain, personal communication, 1999) seem to know when the fence is on or off. While the solenoids in the electric fence control box emit quite loud clicks when the fence is on, these clicks cannot be heard over long distances away from the box. Elephants nevertheless detect whether the fence is on or not. Whether they do this by detecting the electric field or touching vegetation is not known.

When the electric fence is on, elephants have been seen to break sizable tree limbs or branches, which they have then thrown onto the fence and effectively disabled the fence by shorting it to the ground. Larger females have been seen to push younger elephants into the fence to achieve the same ends. Alternatively, a mother elephant was seen to twist the conducting wire around her tusk and break it to let her calf through. In the first instance, the tree limbs are being effectively used as tools to disable the electric fence. In the second instance, another elephant is employed as perhaps the only, if not the most readily, available tool. Likewise, the tusk is being employed as a tool.

Hot chilies have been used as a barrier to protect crops from elephants in various locations. As reported by Edwin Mbulo (*Sunday Post* Online, 29 January 2012), elephants in the Mandia region along the Zambezi River soon learned to deal with the chili barrier. Cloth strips were soaked in chili extract and hung around the perimeter fences of the crops the villagers wished to protect from the elephant raiders. The elephants not only used branches to throw onto the fences and bring down the chili strips, but walked backward across the downed barrier, presumably to avoid getting chili on sensitive trunk tips, eyes, and mouth.

Anthony and Spence in their book *The Elephant Whisperer* (2009, p. 242) relates the capture of 12 Nyala antelope (*Tragelaphus angasi*) (Figure 11.2). The male Nyala are a beautiful black, golden brown, and white striped antelope with 30-in. lyre-shaped horns, weighing in at between 250 and 300 lbs and standing nearly 1.3 m (4 ft) at the shoulder. The females are smaller, golden

**FIGURE 11.2**    Family of Nyala (*Tragelaphus angasi*) with one male and four females. *Photograph courtesy of author.*

brown with vertical white stripes, and no horns. The species has been decimated everywhere in South Africa except the Natal region where Anthony's "Thula Thula" game farm was located. He had set up a capture of Nyala for transloca-tion to one of their old habitats. The capture had been successful and 12 Nyala were safely fenced in an impregnable thorn bush boma. The capture team was seated around their campfire next to the boma enjoying their last night of a suc-cessful capture operation. Then they heard the sound of Thula Thula's herd of elephants led by the matriarch Nana. The team gave way thinking that the el-ephants had been attracted by the smell of the bales of alfalfa that they had been using to feed the Nyala. Clearly, if the elephants wanted the alfalfa they could have it. Instead, the herd stopped as if on instruction. Nana walked deliberately and alone to the boma gate, which was not locked but secured only by closing the hasps on the u-bolts. She manipulated the hasps with the tip of her trunk, first getting one then the other open. Then with her trunk she pulled the gate open, stood aside, and waited. After a few seconds the Nyala responded, found the opening, and were gone. As the last Nyala disappeared, Nana went back to the herd and led them away. They showed no interest in the alfalfa, leaving this prime delicacy untouched.

Nana was once a captive elephant herself penned in a boma. She had been translocated from near the Kruger National Park to Thula Thula. Whether her past influenced her behavior or not, there is little doubt that she was aware of the capture operation, which had used helicopters to round up the Nyala. She would have memories of such events. She was also no doubt aware of the dis-tress calls of the Nyala both during capture and after they were penned up in the boma. Not only was this an act of empathy and altruism on the part of Nana on behalf of another species, but it involved a number of intelligent actions that would qualify under the definition given earlier. It is also of note that both Nana and her herd would have been fully aware of the proximity of humans, their fire, sounds, and pervasive scent. These signals of potential danger were totally ignored and the whole operation carried out without any apparent reaction to the presence of humans.

Elephants have been credited with making tools (Byrne and Bates, 2011b). They use branches to scratch and remove parasites (ticks) from places on their body that are difficult to reach by using scratching posts or old termite mounds. They will even use probes made of small branches to delicately penetrate the temporal gland or the ear orifice. Elephants will spray water over their whole bodies and then blow dust onto the wet skin surfaces (Figure 11.3). Alternatively, they will wallow in mudholes or spray a slurry of mud over their bodies. They will consciously and with considerable vigor stir up the mud to make this slurry. In either event, the mud will dry, protecting them from the sun, potentially cooling them and encasing parasites. By rubbing or scratching the caked mud, parasites will be removed (Figure 11.4). Elephants frequently have a favorite stump or old termite mound, which everyone uses as a scratching post. Hard-to-reach places or cracks and folds in their thick skin, not accessible using a scratching post, are reached with sticks that are broken at the appropriate length. They will go through a trial-and-error routine to arrive at a tool that will do the job at hand.

Elephants will also use branches with leaves or fonds as fans to cool themselves during the very hot periods of the day. Elephants recognize foreign objects such as tranquilizing darts on others and remove them, even wiping injuries

**FIGURE 11.3**  Termite mounds, consisting of fine-grained soil processed by the termites, will be tusked up and converted into dust, which will then be blown over their previously wetted bodies.

**FIGURE 11.4**    Dusting at a waterhole.

with clumps of grass manipulated with their trunks. Using sticks to get to food that is out of reach, piling vegetation to block access, or waving branches at vehicles to repel them, including throwing missiles with considerable accuracy and force, are all within their repertoire of using tools.

Water is stored in the pharangeal pouch of the larynx. Elephants will reach down into this pouch with their trunks and selectively wet their ears to induce added evaporative cooling under extreme conditions of heat. They will also use their tusks as tools to dig minerals from soil banks or as levers to break grass and other vegetation. Tusks will be used to pry bark lose from trees, which is then stripped off the tree with the trunk.

They have also been known to spray both keepers and spectators with water. This has occurred in the wild where Ranger Louis Olivier encountered a bull elephant drinking in the Nwaswitsontso River in the Kruger National Park (Roderigues, *The Game Rangers*, 1992, pp. 38–39). An adult elephant can hold some 8–101 (2–2.5 gal) of water in its trunk. In this instance, the elephant deliberately filled its trunk with water and, employing subtle tactics of deception pretending that it was not aware of Olivier's presence, came within three paces of him, then let him have it with a full blast from the straightened trunk pointed directly at him.

Bates and colleagues (2007) used three identical red cloths in Amboseli, Kenya, differing only in smell. One cloth was impregnated with smells from Maasai warriors, one with smells from Kamba agriculturists, and the remaining cloth was left odor-free. The Maasai, whose land is encompassed in the Amboseli Game Reserve and who graze their animal herds in the park, not infrequently clash with the resident elephants. Under drought conditions, water and food resources diminish and clashes including spearing of elephants occur. The elephant herds in Amboseli are thus wary of the Maasai and will respond to them by flight and occasionally aggression. The Kamba are agriculturists and live in relative harmony with the Amboseli elephants. The elephants show no fear of the Kamba.

When Bates and her colleagues displayed the three cloths to these elephants, only the red cloth impregnated with the odor of the Maasai triggered fear and flight. The other two red cloths were ignored. The experiment was repeated with two clean cloths, one white and one red. The elephants reacted only to the red cloth, showing both fear (flight) and aggression. These reactions of the Amboseli elephants show that they were able to subdivide another species into two classes, demonstrating sophisticated discriminatory and classification abilities based on both smell and sight.

Reports of the early use of aircraft in Kenya to selectively hunt elephants with large tusks suggested that these elephants soon learned to recognize aircraft as dangerous and reacted by disappearing into forested areas. In recent times, many elephants have been translocated after being subjected to traumatic capture and transport. These elephants, often raised without any normal social interaction with other elephants, do not always exhibit obvious abhorrent behavior as described earlier for the Pilanesberg young males. Instead, however, there is evidence of distrust or even dislike of humans. In Hluhluwe Game Reserve in KwaZulu-Natal where such translocation of elephants has taken place, elephants have attacked visitors in cars, displaying possible guile where one group has blocked the road while another group has circled back through the bush and attacked the cars at the end of the stopped line of vehicles.

Social interactions such as described earlier require the knowledge of individual identities and considerable use of cognitive powers. In addition to using sound and communication to identify their own species at distances well removed from themselves, elephants have been shown to use olfactory cues to identify others. Results from experiments conducted in Amboseli National Park in Kenya by Bates and her colleagues (2008b) were able to show that elephants can probably keep track of all of the individuals in a group of as many as 17 elephants. They know where each individual is and can keep track of where these individuals are going, as well as their previous locations. While other cues may also be used, such as sound, Bates could show that smell alone provided the needed information.

Similarly, African elephants have been shown to form clear coalitions and alliances that involve reciprocity and cooperation (Byrne and Bates, 2009). Elephants will attempt to help those that they perceive as being in distress. When elephants have been tranquilized for various reasons, it is often difficult to prevent others in the family group from coming to the assistance of the darted animal, some attacking and chasing away human team members, others trying to assist the elephant that is down by raising it with their tusks and trunks. They are clearly aware of unusual circumstances and that individuals are in distress. While their actions under these circumstances have been observed many times, their vocal response has not been recorded and examined. It is likely that vocalization under such circumstances ranges over a wide band of frequencies, significant parts of which are in the infrasonic range below the threshold of human hearing.

Elephants detect, have knowledge of, and react to a wide range of abiotic sounds such as helicopter rotor noise, gunfire, vehicles, aircraft, seismic noise, and other possible sources of infrasound such as low-frequency sounds generated by thunderstorms.

In an exercise near the Okaukuejo Tourist Camp in the Etosha National Park (ENP) in Namibia where rifle fire occurred about 5 km (3 mile) from the camp, a herd of elephants led by a matriarch had been drinking at the camp waterhole. Just before the gunfire occurred (not audible to humans at the waterhole), the matriarch had given the "let's go" rumble and started to lead the herd off in a direction toward the gunfire. At the time of the gunfire, she stopped the herd and returned them to the waterhole. The herd, having drunk their fill as well as bathed, were confused by the matriarch's action, but she held them there milling around, undecided what to do. A while after the gunfire had ceased, she once again signaled to go and led the herd in the opposite direction from her original proposed course.

Elephants in the Etosha National Park are subject to anthrax, which is endemic to the area (Conrad Brain, personal communication, 1999). Mortality from anthrax is thought to keep elephant numbers in check in the park, alleviating the need for other methods of population control. Only one culling operation has ever been conducted in the ENP so that elephants in the park are not familiar with gunfire. Yet this matriarch recognized these sounds from 5 km (3 mile) away, knew them to represent danger, and took avoiding action.

Other than humans, lions and hyenas are the only predators that pose a threat to elephants. It takes a group of perhaps 10 or more female lions or hyenas to successfully kill an elephant that is part of a herd. Male lions, weighing over 400 lbs each and twice as heavy as females, are powerful enough to single-handedly kill an elephant. More typically two full-grown male lions will work in tandem to bring an elephant down. Although those tuned to the African bush are capable of distinguishing between the roars of male and female lions, it has taken researchers until recently to develop technology that can make such a distinction. McComb and colleagues (McComb et al., 2011), using recordings of male and female roars together with computer software, were able to show distinct differences between male and female lion roars. When recorded lion roars from one to three males or separately from groups of female lions were played to some 39 groups of elephants at Amboseli National Park in Kenya, they were able to show that elephants responded most strongly to the near simultaneous roars of three male lions, less so to a single male roaring, and significantly less to the roars of the lionesses. Furthermore, it was the older (60 years or more) matriarchs who showed the most response not only in reacting herself but in drawing the herd into defensive posture or acting quickly and aggressively to the perceived threat. The older matriarchs made clear distinctions in their reactions to the roars of the males versus the females, demonstrating the ability to discern between different known levels of danger. These findings reinforce earlier descriptions of older matriarchs carrying accumulated knowledge and cognitive ability to interpret threats to their herds and take actions to offset these threats. The author, seated in an open Land Rover, experienced the roaring of two large male lions lying in the grass no more than 3 m (10 ft) away. The sound-pressure level of these roars was sufficient to cause the sheet metal panels of the vehicle to vibrate. An estimate of the loudness of these calls was that they were close to 120 dB. Garstang et al. (1995), reporting on observations made by Stander and Stander (1988), pointed out that lions roar almost exclusively at night and most often near sunset and dawn when the strength of nocturnal inversions is greatest and the sound of such roars can probably be heard by other lions over distances of at least 10 km (6 mile) or an area of over 300 km$^2$ (118 mile$^2$).

Navigation and the elephant's detailed spatial knowledge over wide geographic areas not only include the ability to remember specific locations but have a sense of time such that a specific location is visited only when the food located there is ripe. There is further evidence that elephants will adjust the timing of their visits to such locations based on what the weather has been. Further evidence, described earlier, suggests not only that elephants can detect infrasound generated by thunderstorms but that they know that these sounds mean water and potentially food. They may also have a sense of the drainage patterns of the terrain, and know that rain in catchment areas of ephemeral rivers means water downstream in what have been dry riverbeds over months if not years. In response, they do not expend energy ascending into the catchment region where the rain occurred but head to the nearest downstream location of the dry river channel to await the pending flood.

Behavioral responses of the African honey badger or ratel (*Mellivora capensis*), a species unrelated to the elephant, may help penetrate the enigma of animal intelligence.

The ratel, in common with the elephant, spends an extended period of time under the tutelage of its mother. Researchers have only recently discovered that what were thought to be badger pairs were, in fact, the mother and her offspring, staying together for as much as 2 years (BBC Nature, PBS, June 2014, http:// youtu.be/c36UNSoJenI). Such extended mother–offspring contact suggests the need for significant learning, compounding the issue of distinguishing between learning and intelligence.

At his rehabilitation facility on the western edge of the Kruger National Park in South Africa, Bryan Jones has developed great respect for what he unquestionably believes to be the intelligence of the honey badger (PBS Nature, June 2014). A particular badger named Stoffel by Jones proved to be an incorrigible escape artist. Jones eventually decided to build a 6-ft-high concrete wall surrounding Stoffel's enclosure. Stoffel escaped from this prison the very first night by climbing a tree near the wall and bending the branch he was on to meet the wall. Jones removed all the trees near the walls of the enclosure, only to find Stoffel breaking a branch from a remaining tree, dragging it to a corner, propping it up, and climbing out.

All trees in the enclosure were then removed, only to find that Stoffel dug up stones in the enclosure and piled them in the corner and once again outwitted Jones and escaped. Removing the stones, Jones left a rake and shovel in the enclosure. On successive nights Stoffel used first the rake and then the spade propped up in the corner as a means to escape. Recall that the adult honey badger weighs in at about 25 lbs, is a stocky animal with short legs and very strong claws, standing about 25 cm (10 in.) at the shoulder with a body length of about 46 cm (18 in.). He is not designed to carry or drag long objects such as branches or garden tools nor to carry rocks. Incidently, when all rocks were moved, Stoffel used the dug-up ground and available trough of water to make mud balls, which he then piled in the corner in place of the rocks. The point of this story is to emphasize the fact that when an animal is faced with a problem not of our creation or design but confronting him or her, the intelligence displayed is remarkable.

Elephants in their natural environment have faced and solved innumerable problems of which we, as modern humans, have little knowledge or perception.

Chapter 12

# Learning and Teaching

How much animals learn, how much of their behavior is instinctual, and how much of this learning is by imitation remains controversial. There is no consensus among neuropsychologists and others that any animal, including nonhuman primates, are able to teach or learn through explicit instruction (Caro and Hauser, 1992; Thornton and Raihani, 2008). We have suggested that a large part of human behavior, perhaps as much as 90%, is controlled by the unconscious, leaving only 10% in the conscious realm. Yet we conceded that at some point what may now be unconscious was consolidated through the conscious and through learning.

Scientists generally agree that for humans, language and communication involve a high degree of teaching and learning. Yet there is a reluctance among ethologists to accept that nonhuman animals transfer information through communication. Dawkins and Krebs (1978, p. 308), quoted in Seyfarth et al. (2010), claim that "it is probably better to abandon the concept of information transfer altogether," apparently agreeing with Owings and Morton (1997, 1998) that to do so is both anthropomorphic and inaccurate. Seyfarth et al. conclude that animal calls do contain information that affects the receiver's responses and can illicit more than one response and that "animal signals encode a surprisingly rich amount of information" (p. 7). They arrive at these conclusions without discussing any of the findings on elephant communication and with no recognition that much of that communication is in the infrasonic range, not audible to humans, and in the broader sense poorly observed and understood.

Caro and Hauser (1992), together with Thornton and Raihani (2008), believe that teaching must be explicit, although not dependent upon, one teacher and one learner and must result in the pupil acquiring new skills or competence. While acknowledging the role of mothers and social learning, they do not consider this to be a true form of teaching. Surprisingly, they include no discussion of play as a form of teaching or learning. Teaching, however, should result in learning a new skill or acquiring new knowledge. Teaching should result in changes in behavior and the acquisition of new skills.

Much of elephant learning must be through observing group members, with individuals spending an entire lifetime within a single group continuously following older more knowledgeable relatives. The nearly 2 years that a calf nurses followed by 8–10 years of close contact with the mother and attention paid

Elephant Sense and Sensibility. http://dx.doi.org/10.1016/B978-0-12-802217-7.00012-0

**91**

by allomothers to the calf provide an extended period of learning at a critical time of development for the calf. Almost uniquely among mammals, female elephants remain within a closely knit family unit for the rest of their lives. Behavior learned within this group consists not only of the crucial knowledge needed for survival but equally important knowledge involving social skills and interactions with their family and a much wider population that may number well over 100 individuals.

In contrast, males leave the herd at puberty and show distinctly different upbringing when compared to females. For example, crop-raiding is a high-risk, high-reward exercise that is carried out mostly by male elephants. Energy requirements of males rise at the age of first reproduction (25–30 years old). At their reproductive peak (45–50 years), males' nutritional demands are at the highest level. Moss and her colleagues (Chiyo et al., 2011) have shown that males at their prime take more risks than younger males or females. Females in the Amboseli herds, although exposed to crops, do not take the risks entailed in crop-raiding. Younger males, however, join with experienced older males, suggesting that they have found that social learning is better than solitary learning, especially when the cost of exploratory learning may be high.

Failure to observe teaching in the case of elephants may be a function of the lack of opportunity to observe such a process rather than the conclusion that teaching among elephants does not exist. A more reasonable conclusion may be that explicit teaching is not a frequent function in elephant society and as such rarely observed. A potentially explicit case of teaching by an elephant mother of its calf was observed in a dry riverbed in the Kruger National Park by the author (Figure 12.1). The mature female, possibly the matriarch of a small herd, had successfully dug for water in the sand of a dry riverbed. Her calf, perhaps 4–5 years old, was obviously interested in the water at the bottom of a 1 m (3 ft) deep hole. However, it neither knew how to drink using its trunk nor how to reach the water at the bottom of the hole. The mother took up a quantity of water in her trunk. She wound her trunk around that of her calf such that the tip of her trunk was immediately below the tip of her calf's trunk. She then allowed the water in her trunk to flow out such that it welled up into the tip of the calf's trunk. It seemed obvious to the observer that she was trying to get the calf to take up the water in its trunk. The calf spluttered and struggled but failed to get the idea. The mother tried repeatedly to get the calf to respond but after about four or five tries gave up, took a trunk full of water and pumped it directly into the calf's mouth. From the sequence of actions taken by the mother it seemed that she had the clear intent of getting the calf to use its trunk as a means to drink. She clearly knew that the calf had to take up the water in the lower end of its trunk and transfer this water to its mouth. She did everything she knew to show the calf how to do this. When this failed she knew that the transfer process from the trunk to the mouth by the calf was not going to work and that she had no option but to execute this part of the operation herself. Teaching may thus be an uncertain process that needs to be repeated many times before the lesson is

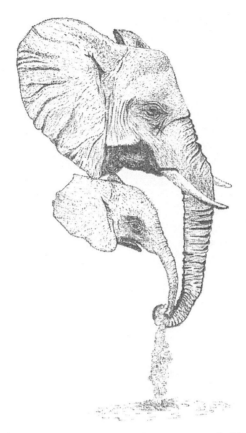

**FIGURE 12.1**    Mother repeatedly tried to get the calf to take water she had drawn from a hole dug in the river bed too deep for the calf to reach. (Pen and ink drawing by the author.)

learned. It is possible that elephants perform such repetitive instruction teaching young by multiple subtle demonstrations.

Earlier we described a lone female African forest elephant who seldom if ever vocalized except when apparently hearing distant low-frequency sounds generated by the wake vortices from a 747 Boeing aircraft taking off from the Lahore airport. While there is no evidence that this elephant mimics the sounds she heard, more recent observations have documented that elephants can learn to imitate the sounds of truck engines (Poole et al., 2006). Stoeger et al. (2012) has documented that male African elephants living with two female Asian elephants have learned to mimic the chirping sounds made by the Asian elephants. Stoeger also documents an Asian elephant mimicking human speech.

Vocal learning not only means that elephants are aware of sounds made by other animals and nonanimal sources but that they have the potential of developing an open communication system that potentially could include humans. Byrne and Bates (2009, p. 72) suggest that there seems to be some teaching behavior inherent in the attention paid by older females to a young female when she first comes into estrous.

de Waal (*The Bonoko and the Atheist*, 2013, pp. 114–115) describes an experiment conducted at the Washington National Zoo by Preston Foerder and Deana Reiss. Kandula, a young elephant bull, was presented with bunches of fruit being suspended out of reach from the roof of the enclosure. Several objects, including sticks, a sturdy box, and several thick cutting boards, were first distributed around the enclosure. Kandula ignored the sticks but after a while moved the box with his foot in a straight line until it was under a bunch of fruit. He then stood with his front feet on the box, reaching up to the fruit with his trunk. Foerder and Reiss then moved the box outside and out of sight from the enclosure. Without hesitation, Kandula retrieved the box, from apparently considerable distance, bringing it once again to a position below a bunch of fruit. They then further complicated the experiment by removing the box entirely, leaving the sticks and wooden boards. Again Kandula ignored the sticks, picked up the wooden boards, and stacked them on top of each other to serve as a platform to reach the fruit.

While elephants commonly use their trunks to reach high into trees for fruit and other edible plant material, and will make use of terrain, roots, and other small elevated surfaces to add to their reach, they have not been observed in the wild to move objects to stand on (Figures 12.2 and 12.3). In the Washington Zoo case, Kandula showed not only deductive reasoning but a learning process that he could recall, repeat, and substitute even when the means (tools) were no longer visible or available. While this experiment was conducted under artificial conditions, it had elements based on natural conditions that utilized both the

**FIGURE 12.2**  Foreign objects are readily recognized for what they are: a potential source of clean water if broken into, which they often are.

**FIGURE 12.3**   Reaching for objects perhaps as high as 8 m (25 ft) above the ground. (Pen and ink drawing by the author.)

neural and physical assets of the elephant. Being able to either observe behavior in the natural world or effectively transfer natural simulations to artificial environments is essential in trying to assess the intelligence of animals.

Sitting behind a desk is a dangerous place from which to view the world (John le Carré quote; http://www.goodreads.com).

Chapter 13

# The Sensory Environment
# of Elephants

Sensory signals are transmitted in multiple ways, consciously and unconsciously, by individuals and through group responses. Individuals transmit
signals through the five sensory systems. Conscious signals are frequently obvious. My nephew, Richard Garstang, and I demonstrated the latter when being harassed by a young, fairly aggressive bull elephant in the Kruger National
Park. The young male had repeatedly mock-charged our Volkswagen "Combie."
After about the third episode we waited for another frontal threat, then as the
youngster came toward us we simultaneously swung open the front doors of
the Combie in a realistic imitation of the flaring of the ears at an adversary. The
feisty youngster skidded to a halt, turned, and abandoned his game in immediate
response to a well-understood threat from a larger opponent.

Unconscious signals are less obvious, more subtle, and not under the conscious control of the animal. Pheromone emitted signaling fear or aggression
are not controlled. Other subtle signals in tone and body language may not be
consciously generated (Figures 13.1–13.4).

Collective signals may be generated by groups and large assemblies.
Humans, for example, believe that in sports events crowd energy can be transferred to players. This may well be achieved in a number of ways by sound,
smell, and collective body language. These responses have evolved over evolutionary time, particularly in social and herd animals, and may be far more
prevalent in animals than in humans. Ecologists and behavioral biologists have
focused on conspecific communication and some interspecies communication
but paid scant attention to overall environmental acoustic, olfactory, visual, tactile, and gustatory input.

Of the sensory input, the auditory component may be the most transform-
ing and may be far more pervasive than has been previously considered. The
sensory surroundings of an elephant consist initially of its mother's womb and
ultimately of its total environment. Detectable initially by touch and smell and
subsequently by sound, taste, and sight, it ultimately involves all five senses,
which serve to identify the elephant's place in its world.

Elephant Sense and Sensibility. http://dx.doi.org/10.1016/B978-0-12-802217-7.00013-2

**FIGURE 13.1**    Awareness and tolerance of others is exhibited by clear body language from mild curiosity to explicit warning signs.

**FIGURE 13.2**    Awareness and mild curiosity with no threat.

**FIGURE 13.3** Awareness with some fear and potential flight.

**FIGURE 13.4** Awareness coupled with mild flight.

The sensory systems of mammals (auditory, olfactory, visual, tactile, and gustatory) provide parallel and continuous input from the environment to the brain. Neural systems must function to interpret the signals and assign meaning to the integrated input. Interpretation must trigger memory and response to a large range of situations and images that are recalled. Many such memories are of responses upon which survival may have depended.

The acoustic fields in the natural environment may represent signals that travel over the longest distances and that constitute the basis for mammals such as elephants to recognize spatial and other characteristics of their surroundings.

There is ample evidence that suggests that elephants can travel over large distances (hundreds of kilometers) to reach specific locations at specific times. There is further evidence that animals who have been translocated over similar long distances (in closed vehicles) are subsequently able to return on foot to their place of origin. Orphan elephants at the David Shepherd Wildlife Trust are taken by road in closed vehicles after as much as 10 years in the Trust to be released in the Tsavo National Park approximately 150 km (93 mile) away. Female released elephants have successfully integrated into the wild herds of Tsavo, some of them ultimately producing calves of their own. In a number of instances mothers have brought these calves over a distance of at least 150 km (93 mile) back to the Trust. Quite apart from any attempt to understand the motivation of the mother to bring their calf back to the Trust, no plausible explanation can be offered to explain how the mother knew where the Trust was relative to Tsavo and how she was able to get there. Seeking a rational explanation one is forced to call upon more than one sophisticated use of navigation, memory, and neural competence.

In the case of Lawrence Anthony's death, described in Chapter 7, the postulated response of the herds is to the behavior or, perhaps, more importantly is to the changed behavior of another species. Not only might the elephants have detected this change in behavior but they may have correctly interpreted it as displaying grief.

It is likely that individual elephants have a total neural memory of the sensory environment of their home territory. A significant part of this neural memory consists of auditory sounds stemming from other than conspecifics including humans. Abiotic sounds add to the input. This wide range of sounds includes frequencies and levels that are inaudible to humans. As such, these sounds have been poorly recorded and inadequately studied. While the acoustic environment may extend over the greatest area, it remains a part of the total sensory environment that defines the animal's home territory. This total sensory environment is made up of all of the sounds, smells, sights, and tactile and taste sensations that have been recorded and consolidated within the animal's brain over the time spent in its territory. The sum total of this sensory input serves to provide the basis for the animal not only to instantaneously recognize its home territory but to be conscious of any changes or inconsistencies in the environment of that territory.

The embedded memory consisting of integrated multiple sensory inputs serves as a basis to recognize familiar territory such as the component parts of the home range. It also serves as a neural framework against which changes or unusual circumstances can be recognized.

Recognizing that elephants have such capabilities presents the challenge of interpretation by humans of such a complex synthesis of input but will also have important implications to management and conservation. Inherent in the proposition is the concept that the neural connection to a given environment, which is far more complex than simply a connection to the biota of that environment, is fundamental to these animals. Changes such as translocations represent in these terms fundamental dislocation.

# Chapter 14

# Them and Us

At this point in our effort to penetrate an elephant's mind we need to ask what have we learned about an elephant's mind and what separates them from us?

Despite taking different but parallel evolutionary pathways, elephant and human cognitive systems are more similar than they are different. We each have five senses, humans responding more to sight and sound and elephants more to sound and smell. While humans have progressively lost their dependence upon smell, touch, and taste, we have not abandoned these senses. Elephants, on the other hand, depend on all five senses and may do much better than humans at integrating and assimilating input from multiple sensors, extracting information critical to survival, and storing that information in long-term but accessible memory.

The evolution of both the human and the elephant brain has resulted in substantial storage within the unconscious. Much of the complex functioning of our respective bodies is carried out by the unconscious and both humans and elephants are capable of involved motor skills directed entirely by the unconscious. Despite this ascendancy of the unconscious in both species, the unconscious is ultimately dependent on the conscious and there is continuous feedback between the unconscious and conscious.

Memory is vital to the well-being of both species and memory in both species evolved within a spatial framework of the surrounding world. Survival depends critically on knowing where food, water, shelter, and danger lie. This spatial knowledge had to be precise and not independent of time. Spatial memory supported survival and memory played a key role in the social systems of both species. Both species depend on and are part of a highly complex social system. In the case of elephants, males are separated from females but the raising of the young is prolonged and intimately entwined within the social system. Protracted care of the young and comparable long life within stable societies meant the evolution of rules of behavior or morality and the emergence of empathy, altruism, and emotions. Thus, elephants recognize self and others, form long-term bonds and coalitions, and take part in cooperation. They display a wide range of communication skills that includes all of their sensory systems, resulting in measurable intelligence, strong communication skills, and perhaps even rudimentary language and the ability to teach others. They display the ultimate in the recognition of others and self in their nearly unique response in

Elephant Sense and Sensibility. http://dx.doi.org/10.1016/B978-0-12-802217-7.00014-4

the nonhuman animal world to death and dying. Elephants are sentient beings differing from humans only in degree and not in kind.

Such conclusions certainly require that we reexamine the relationship between ourselves and elephants (and perhaps all other animals). If we as humans have fundamental behavioral characteristics relative to elephants that are more in common than they are divergent, what sets us apart and what justifies the acceptance of significantly different privileges for one species versus the other?

Elephants were present on the savannas of Africa some 12 million years before the upright walking hominid diverged from the apes. *Australopithecus afarensis* as represented by Lucy (named from the Beatles' song "Lucy in the Sky with Diamonds"; Johanson and Edgar, 1996, p. 37) appeared between 3.9 and 4.4 million years ago (MYA), perhaps overlapping with later hominids (*H. habilis*, *H. ergaster*, and *H. rudolfensis*). The fairly sophisticated "Acheulean" hand ax emerged 1.5 MYA, possibly coincident with a primitive form of language. By 500,000 years ago (YA), *Homo heidelbergensis* were regularly using fire and building shelters, although there are few fossils and hence little evidence of this age in Africa.

There was sudden change in the sophistication of stone tools in the so-called Middle Stone Age about 250,000 YA. Tools became hafted and axes and spears appeared. Modern humans in the form of *Homo sapiens* (wise man) in Africa may date from as early as 200,000 YA.

The preceding apes and the succeeding early *Homo* were not large in either weight or stature. During the australopithecine phase of human evolution, males weighed between 40 and 50 kg (88 and 110 lbs) and stood between 1.3 and 1.5 m (4.2 and 4.8 ft) tall, females weighed between 28 and 34 kg (60 and 75 lbs) and stood between 1.1 and 1.2 m (3.5 and 4.2 ft) tall. A significant increase in size did not occur until the emergence of *Homo ergaster*, some 1.5 MYA based on the findings of the Turkana Boy. This nearly complete skeleton yields a weight of 67 kg (144 lbs) and height of 1.6 m (5.1 ft). Considerable variability, however, must have existed, for a nearly contemporary *Homo habilis* skeleton was found and estimated to weigh only 24 kg (52.8 lbs) (Johanson and Edgar, 1996, pp. 57–73).

Small in stature and numbers, these hominids would have presented little threat to elephants until they had developed hafted spears, possibly fire, the ability to dig deep, wide pits, or perhaps find and use lethal poisons. They must have preyed on and scavenged sick, dying, and dead elephants, and would have capitalized upon elephants in distress in mudholes or in other dire situations. It is unlikely, however, that elephants regarded these early humans as a serious threat. These small hominids would more likely have avoided contact or direct confrontation with elephants. It is unlikely that they encroached willingly onto the open savannas and would have certainly avoided being found on the open plains at night. Instead, the remaining riverine forests, which, in turn, had water and rocky prominences with caves and sheltering ledges and overhangs, would have been preferred areas. These are the locations where the presence of early hominids have been found by paleoanthropologists such as Dart, Broom, Leaky, and others. The caves and shelters of the San or bushmen reflect the probable way of life of these early African hominids.

Humans must have viewed elephants with awe, fear, and even reverence. Tribes in central Africa (Acholi people of Uganda; Bradshaw, 2009) and the Zulu of Natal (Mutwa, 1965, p. 486) have the elephant as their totem. Oral tradition of the Ndorobo of Kenya tells of a time when elephants and people coexisted and shared resources including elephant milk for Ndorobo children (Christo and Wilkinson, 2009, p. 13). Ganesha, the elephant God in Hindu religion, has been revered for centuries. Buddhists of Tibet show similar reverence of elephants.

It was not until modern times with the advent of primitive firearms and the appearance of Arab and European slave and ivory traders that humans in Africa began to threaten elephants. For the greater part of their evolution, elephants must have had little fear of their fellow animals, and no particular fear of the human species.

This relationship and, in particular, that between human and elephant as opposed to elephant and human was, for millennia, benign if not reverential. Such a relationship, however, was seated between people and elephants in Africa and in parts of Asia, but not in Europe and much of the Near East.

Although the mammoth had been encountered and perhaps even exterminated by *Homo sapiens* in Europe, insignificant contact between Europeans and elephants existed prior to the penetration of Africa by Europeans after the turn of the 19th century. Up to this time in Europe, elephants were not seen for what they are but as exotic and foreign curiosities progressively embellished with human trappings and estranged from their natural habitats. In the United States elephants were seen as objects to be exhibited and displayed for commercial purposes.

By the 19th century, Dutch and English settlers had occupied a large part of the southern tip of Africa. Elephants in the Knysna forests of the Cape Colony were being systematically destroyed and ivory hunters were penetrating the interior from the south and beyond the Zambezi. Portuguese traders and slavers had established footholds in Mozambique (Lourenco Marques and Beira) and Arabs were well established in Zanzibar and on the African east coast. The mindset brought by these outsiders to Africa was one of exploitation and plunder. Like the influx of Europeans into South and North America, they demonstrated unbelievable disregard for the abundance of new and unfamiliar wildlife. In total contrast to the views held by indigenous peoples of the Americas, Europeans managed, in the space of a few generations, to obliterate entire animal species (passenger pigeons), and bring to near extinction many others (bison, wolves, bears). In the 3–4 years between 1870 and 1873, large scale slaughter of the bison on the high plains of the American West had claimed over 5 million animals. A hunter named Tom Nixon shot 120 animals in 40 min. In 1873, he killed 3200 bison in 35 days (Gwynne, 2010). Once this onslaught was brought to Africa, the relationship between humans and elephants in Africa changed dramatically.

It is possible that there is a deep-seated memory within elephants that still views humans as benign and nonthreatening. Despite the treatment meted out by humans to elephants, there are still many instances of bonding and trust between them. The two or three centuries of violence may have been sublimated by these animals wherever conflict is removed. If we believe that evolution has molded the complex systems running the elephant's brain, then it is possible that the events of a few centuries have not overwhelmed the neural foundations laid down over millions of years of evolution. Evidence that has been accumulated by long-term studies of elephants, such as in Amboseli by Cynthia Moss and Joyce Poole, and elsewhere by Iain Douglas-Hamilton, Dame Edith Sheldrich, and others, has suggested that when elephants are not threatened by humans they exhibit remarkable tolerance for them. Conflict between elephants and humans occurs when elephant habitat is lost, elephant movement is restricted, and extreme measures are taken by humans to pursue their needs at the expense of those of the elephants.

Where humane methods have been assiduously applied, such as in the rescue of orphaned elephants at the David Sheldrich Wildlife Trust, remarkable reciprocity between elephants and the human caretakers seems to occur. As described earlier, Dame Sheldrich has documented cases where elephants arriving traumatized by their experience of losing their mother and entire family have been brought to adulthood and released into the wild (Tsavo National Park), where they have successfully mated and borne offspring. They have then returned over 161 km (100 mile) to the shelter together with their calves. However reluctant scientists may be draw conclusions from such anecdotal evidence, it behooves us as humans to take note of the possibility that this behavior on the part of the elephant rests upon a deep-seated knowledge of humans and the relationship between humans and elephants.

Whatever relationship existed between people and elephants prior to the influx of Europeans into the African continent after the turn of the 18th century, the treatment of elephants by humans changed dramatically. Explorers turned hunters and ruthless exploiters of the vast herds of animals found in Africa quickly made serious in-roads into these populations and rapidly changed the relationship between humans and animals. By the end of the 19th century, the unreliable and frequently ineffective smooth-bore muzzle loaders had been replaced by powerful and deadly accurate hunting rifles. Early hunters such as R.J. Cunningham, J.A. Hunter, Frederick Courtney Selous, Theodore Roosevelt, and W.D.M. Bell collectively destroyed thousands of elephants. Bell alone is credited with killing 1000 elephants in southern Africa (Bell, *Bell of Africa*, 1960, p. xiii). To this must be added hundreds shot by him elsewhere in Africa. His biographer adds with pride that these 1000 animals were shot by Bell himself and not followers and further that "he probably killed a larger proportion of his beasts with a single shot, and he was never mauled." With the much vaunted code of honor among hunters, Selous and others would shoot a female elephant with a calf or shoot one elephant, only to abandon it in favor of another with larger tusks. When European settlers arrived at the Cape, there were on the order of 100,000 elephants in present-day South Africa. By 1910, there were less than 200 elephants in four fragmented populations (Patterson, 2009, p. xi). By 1994 when Patterson began research on his book *The Secret Elephant*, there was reputedly only one elephant left in the world's southern-most elephant population in the Knysna forest of the Cape Province.

While the destruction of elephants was devastating in the colonial period of the 19th and early 20th centuries, it paled in comparison with what was to come in the post-colonial period of civil wars and political instability in Africa.

Human populations in Africa suffered social disruption on the scale of the elephants. Industrialization in Africa had already disrupted family and tribal life. Fathers were separated from their families and children grew up in rural areas without education, guided only by their unschooled mothers and other female family members. As unrest spread, children were armed with automatic weapons, subjected to brutal treatment, and taught total disregard for both human and animal life and suffering. In the civil war in Angola, as in many other areas of conflict, elephants were slaughtered for their ivory to buy arms and for meat to feed troops. Between 1980 and 1989, 100,000 elephants were killed in this conflict. Modern automatic weapons, together with other technology such as helicopters and all-terrain vehicles, increased the efficiency with which poachers could operate and diminished the capacity of hopelessly outnumbered and underfunded conservation forces to combat the depredations.

In North Luangwa National Park in Zambia, 93% of the elephant population was killed. Traditional elephant herds consisting of close family members were disrupted and young were raised by inexperienced mothers (Bradshaw, 2009, p. 62). In Zimbabwe, it is estimated that a population of 10 million elephants has dwindled to a few hundred thousand but no one really knows. The Zimbabwe government has either allowed or turned a blind eye to marauding bands indiscriminately killing animals and poisoning waterholes in game parks. The situation in Mozambique was as disparate as in Zimbabwe during the conflict in that country. Chaos continues to exist in the Democratic Republic of the Congo and adjacent Burundi.

Exploding human populations in Africa in locations relatively free of open conflict continue to accelerate habitat destruction, encroachment of protected areas, and human–elephant conflict. With habitat loss and human expansion, existing elephant sanctuaries or protected areas are is progressively surrounded and isolated. Ancient migration routes are severed and the ability to utilize resources and to maintain healthy genetic populations is progressively reduced or eliminated. Kenya's human population grew from 8.6 M in 1962 to over 30 M in 2004 while the elephant population plummeted from 167,000 in 1973 to 16,000 in 1989 (Bradshaw, 2009, p. 62). Unless human populations in Africa stabilize in the very near future (next 20–40 years), competition for resources will result in unsustainable reduction in habitat and the ultimate loss of species.

Most critical in the sustainability of all species in Africa is the inherent limitation placed upon survival by the African environment. Seventy percent of the continent of Africa receives less than 50.8 cm (20 in.) of rainfall per year. This amount of rainfall in 1 year represents the boundary between agriculture and pastoralism. Crops cannot be grown reliably without at least 50.8 m (20 in.) of rain per year. Add to this fact that the reliability of rainfall is directly related to annual rainfall. The wetter the region the more reliable the rainfall, the drier the region the more unreliable the rainfall. Africa, outside of 15° of latitude from the equator, is a dry continent with the world's largest deserts: the Sahara, Namib, and Kalahari. Wide areas of the Niger, Sudan, Ethiopia, Eritrea, Somalia, and Kenya are semi-deserts. Much of Namibia, Botswana, parts of Angola, and the Karroo of South Africa are the Southern Hemisphere equivalents. Cycles of drought have been evident in these regions for many thousands of years. Accounts in the Bible of 7 years of famine and 7 years of plenty in Egypt are testimony to this cycle. Many scientists, notably Tyson (1986), have documented the existence of this cycle using existing rainfall records.

Add to the limitations of African climate the lack of both surface and artesian water, together with poor fertility in the bulk of African soils and the constraints imposed by the continent on the number of animals (human as well as others) it is capable of supporting become apparent. In fact, it might be argued that the pre-European distribution of fauna and flora might represent near optimum utilization of that continent's natural resources. Adaptation by animal populations, including elephants to this dry and fluctuating climate, is manifest in their extensive migration routes. It has been these adaptations to climate and food resources that have permitted these animals to survive. Disruption and destruction of the long-established seasonal use of resources will collapse the system.

We as modern humans might do well to recognize what evolution has demonstrated and take heed of that guidance. Instead, as human populations continue to grow, malnutrition and starvation will continue to affect both humans and animals. Without limitations curtailing human populations, the current prospect for the survival of African wildlife is bleak indeed.

As humans have come to recognize the value of natural systems in maintaining a viable world, the complexity and diversity of this world have become increasingly obvious. Seemingly unimportant and even insignificant parts of the system turn out to be critical and the task of conservation grows in complexity. Nevertheless, attempts at conservation were begun with the formation of the first national parks in the late 19th century in Australia and in the United States.

In 1898, President Kruger of the Transvaal Republic proclaimed part of what is now the Kruger National Park (KNP) in South Africa, a national park. Even this seemingly farsighted act of wildlife preservation was not fully motivated by the need to conserve the rapidly disappearing natural world of South Africa. Wildlife in South Africa, as in much of the rest of the world, was viewed as a resource to be exploited. Preservation was pursued as a strategy to maintain wildlife as a resource for future human exploitation. Furthermore, the areas that were set aside for wildlife were by and large not suitable for human habitation at that time. Endemic diseases such as sleeping sickness or in animals, ngana, malaria, and foot and mouth disease, and an extreme climate made many of the early game reserves unsuitable for human occupation.

Loss of habitat due to human encroachment together with European ideas of land ownership and agricultural practices progressively and rapidly displaced wildlife over much of Africa. The limited areas designated as wildlife preserves meant that animals were restricted to a specific area. Once this was done, humans are required to "manage" the animal populations within the designated areas. Management, as scientists and conservationists have learned, is a complex and difficult task. Viewing the area as an integrated whole, where the nature and the strengths of the connections between the living and physical systems occupying the designated space are unknown, was found to be far more difficult than the caretakers ever imagined.

Managing within such an environment of a large social herbivore such as an elephant has presented multiple serious problems, not the least of which has been attempting to decide the number of animals that a given area can support. Once restricted, the population now can exceed the capacity of a given area to support it. If hope is placed on serendipity, then hope can easily be dashed, as happened when the population of elephants in the Tsavo National Park of Kenya was allowed to grow unchecked.

If, on the other hand, wildlife managers attempt to control a population, numerous other difficulties arise. Early attempts at managing elephant populations were disastrous when individual elephants were shot in a herd. Herd dynamics were disrupted, the remaining animals were subjected to extreme stress, and the method recognized as being impossible to carry out. The current alternative strategy has been to kill all of the elephants except the very young in a given family group. Without even considering whether such action would be noted by or affect other elephants in other families or groups, it was not initially believed that the infants saved would suffer ill effects from this horrendous experience.

The routine culling procedures in the KNP were carried out with military precision. Helicopters were used to first find and select the family group of elephants to be targeted. Ground crews in vehicles were positioned relative to the chosen group. These included rangers armed with high-powered magazine rifles, game trackers to assist the rangers, and an entire cadre of butchers to dismember the fallen animals and prepare them for processing in a large abattoir especially built for the purpose. It was emphasized by park authorities that all parts of the elephant were used, the greatest bulk going into canned meat for human and animal consumption. In order to reduce the chance of spoilage, the entire operation was carried out in the late afternoon and early evening when the heat of the day was abating. This was done either in ignorance or in disregard of the fact that at this time of day all sounds from helicopters, vehicles, gunfire, and elephant distress calls, as well as human voices, would be transmitted the greatest distances under the rapidly forming nocturnal temperature inversion. The low-frequency sounds generated by the helicopter rotors, rifle fire, and vehicles would be audible to other elephants over distances of many miles reaching animals in far distant parts of the park and in the many adjacent private game reserves. The screams and frantic calls of both adult elephants and young would be heard by other elephants at least 10 km or 6 mile away, meaning that all elephants within an area of over 300 km$^2$ (112 mile$^2$) would be witness to the slaughter and be traumatized by it. As though one needs to add any more horror to the scene, the young calves that were to be saved were tied with ropes to their dead or dying mothers. Bradshaw (2009), in her book *Elephants on the Edge*, speculates that the trauma experienced by these calves remained with them and manifested itself in abhorrent adult behavior many years later.

While these carefully planned and logically argued operations carried out for the declared purpose of preserving the elephant and in the name of conservation were and, unfortunately, still are being carried out, they represent but a small fraction of the inhumanity of man perpetrated against elephants.

Elephant slaughter carried out by hunters, poachers, and undisciplined paramilitary and military personnel equal and greatly exceed these culling operations.

Direct violence done by humans to elephants has greatly exceeded the horrors of culling operations. The record as recounted by Bradshaw of the treatment of elephants in circuses and zoos in the 18th and 19th centuries has appalling examples. The record of modern-day zoos is only marginally better in demonstrating that any humanity is being extended to these animals. As recently as 1988, Paul Hunter, a San Francisco zookeeper, stated, "You have to motivate them (elephants) and the way you do that is by beating the hell out of them" (Bradshaw, 2009, p. 18). To get elephants to comply to the will of humans, they are starved of food, denied water, kept in isolation, restrained to the point of being immobile, beaten, gouged, and torn with iron bars or metal-tipped instruments. Deprived of contact with other elephants, they lose all sense of normality. For an animal born into a close family group nurtured by sound, smell, touch, taste, and sight and within close contact with the environment, the conditions of even the best forms of captivity are totally debilitating. Bradshaw (2009) cites evidence (p. 104) that in the global captive population of elephants the median life span of African elephants is 16.9 years compared to 56 years in free-ranging Kenyan elephants. Lest one underestimates the number of captive elephants, Bradshaw quotes numbers on the order of 9000 for the globe as a whole. Nicol (2013) puts this number at 5000 based on the Swedish *Elephant Encyclopedia* but points to the additional salient fact that the average number of elephants per holding is less than two, making it obvious that many live without a single companion of their kind.

Elephants subjected to trauma, particularly at a young age, or deprived of care by mothers and close kin as well as social contact, suffer all the manifestations of what is now recognized in humans as posttraumatic stress disorder

(PTSD) (Bradshaw, 2009, p. 112). When subjected to prolonged and repeated trauma, PTSD does not fully describe the condition of the victim. Judith Hennan (Bradshaw, 2009, p. 147) proposed complex PTSD to cover this condition.

It may now be argued, as Bradshaw does, that the cumulative effect of modern humans on elephant populations has been such as to effectively traumatize the entire species. Violent behavior between elephants and between elephants and other species including humans is believed to have occurred at levels unknown in the past. In the Pilanesberg case cited earlier, the young bulls killed 100 rhinoceros. In Addo National Park in South Africa, conflict between translocated adult African male elephants has resulted in deaths approaching 70–90%. In contrast, in the Amboseli National Park in Kenya, Moss reports only four deaths in elephant male-to-male conflicts in 30 years. Young males transferred to the KwaZulu-Natal park of Hluhlwe have been reported to deliberately stalk and attempt to attack tourist vehicles. While generalization to the entire elephant population may not be justified, evidence that there is a link between abhorrent elephant behavior and traumatic dislocation of an elephant's early development is a valid hypothesis. Whatever the nature and magnitude of the cause and effect is, the question of why and how it is possible that the human species can inflict such trauma upon another species remains unanswered.

In confronting this tragic relationship between humans and another species, there are at least two rays of hope: sanctuaries that have worked and an evolutionary history that ultimately may prevail. In 2006, the total population of elephants in Africa was estimated at just under 700,000. Of this number, about half reside in Southern Africa (Angola–Zambia and southward, with the South African population numbering between 17,800 and 18,500) (Scholes and Mennell, 2008) (Figure 14.1).

South Africa, with its checkered history of race relations and apartheid, has received little credit for its effort to recover from a disastrous beginning in wildlife conservation. When what was to become the KNP was first established in 1902 (delayed by the Anglo-Boer War from its proclamation in 1898), almost all elephants had been destroyed in South Africa. Stevenson-Hamilton (*South African Eden*, 1993, p. 191), in his survey of the incipient park, saw only seven bull elephants, which in all probability had crossed into South Africa from Mozambique (Figure 14.2). In the two fragmented areas in the Cape, Addo and Knysna, only a remnant of the southern-most population of African elephants remained. By 2006, there were four viable populations of elephants in South Africa: KNP (12,500), KwaZulu-Natal (1100), Northwest population (760), and Addo Elephant Park (450) (Scholes and Mennell, 2008, p. 23).

The Kruger and Addo populations are growing rapidly and both parks have embarked upon large-scale expansions. The preexisting KNP covers an area of approximately 20,000 km² (7800 mile²) or roughly the size of the State of Massachusetts. The plan now is to add the Parque National do Limpopo lying on the northeastern border of the Kruger to form the Great Limpopo Transfrontier

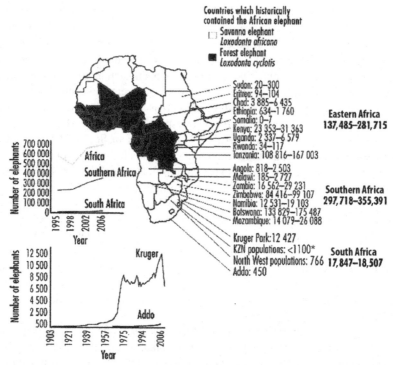

Countries which historically
contained the African elephant

☐ Savanna elephant
   *Loxodonta africana*
■ Forest elephant
   *Loxodonta cyclotis*

Sudan: 20–300
Eritrea: 94–104
Chad: 3 885–6 435
Ethiopia: 634–1 760
Somalia: 0–7
Kenya: 23 353–31 363
Uganda: 2 337–6 579
Rwanda: 34–117
Tanzania: 108 816–167 003

**Eastern Africa**
**137,485–281,715**

Angola: 818–2 503
Malawi: 185–2 727
Zambia: 16 562–29 231
Zimbabwe: 84 416–99 107
Namibia: 12 531–19 103
Botswana: 133 829–175 487
Mozambique: 14 079–26 088

**Southern Africa**
**297,718–355,391**

Kruger Park: 12 427
KZN populations: <1100*
North West populations: 766
Addo: 450

**South Africa**
**17,847–18,507**

**FIGURE 14.1**    Elephant distribution and population trends in Africa. After Scholes and Mennell (2008). *Permission courtesy of the University of the Witwatersrand Press and authors Scholes and Mennell.*

Park, effectively doubling the area of the individual parks and opening up ancient migration routes across the Limpopo (Figure 14.3).

The Addo Elephant National Park is being increased in size more than six-fold to include both mountainous and coastal terrain. The size of the original Addo Elephant Enclosure of about 220 km² (80 mile²) will be increased to approximately 1380 km² (500 mile²), with an additional 128 km² (50 mile²) of coastal parkland and over 1240 km² (450 mile²) of coastal ocean including Stag and Bird Islands. The ultimate goal is to connect the inland park area to the coastal region, allowing elephants to regain access to the sea. Early observations document elephants entering the ocean when access to the sea was open to them (Figure 14.4).

While conflict between elephants and humans has occurred where elephants have been recently translocated, conflict within established populations has been very low. Scholes and Mennell (2008, p. 222, Table 14.1) shows 19 human fatalities over 6 years for all situations in which elephants come into contact with humans. Thirteen of these fatalities have occurred in protected areas, with two of the remaining six in theme parks or elephant-back safaris. Except for newly introduced elephants raised from captured calves, as described earlier,

**FIGURE 14.2**    KNP bordering Mozambique to the east, measuring some 400 km (240 mile) from north to south and averaging about 65 km (40 mile) east to west, roughly the size of the State of Massachusetts. (See also Figure 1.2 in Chapter 1.)

elephants show extreme tolerance of humans in close contact in the South African parks. In fact, the overwhelming impression in watching elephants in these parks is the sense of complete relaxation and absence of any signs of aggression or even concern about human presence. Here the elephant is seen "as a symbol of strength, gentleness, wisdom, and perhaps most importantly, the frailty of the natural world.... In the presence of elephants, the innocent within us is restored" (Christo and Wilkinson, 2009, p. 11).

Humans, in the presence of elephants, almost universally experience a deeply felt awakening of their connection to nature, adding yet another facet to the value of this animal.

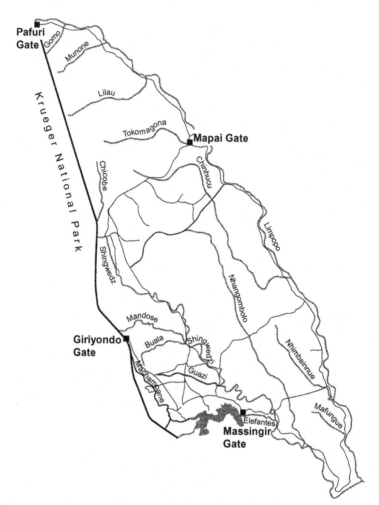

**FIGURE 14.3** Limpopo National Park bordering the KNP, measuring some 200 km (130 mile) north to south and about 90 km (60 mile) east to west. When combined with the KNP, as the Great Limpopo Transfrontier Park, it almost doubles the size of the original reserves. (See also Figure 1.2 in Chapter 1.)

Similar conditions of coexistence exist in Namibia, Botswana, and parts of Kenya. In the Amboseli National Park in Kenya, the Amboseli Elephant Research Project (Moss and Lee, 2011, p. xi) has worked in close contact with elephants for more than 40 years. Despite sharing the park with grazing competitors, the Maasai, conflict between the two is rare and triggered most often when events like extreme drought stress both populations. In the Amboseli-Tsavo Group Ranch Conservation area between 1993 and 2005, 18 people were killed and the same number injured by elephants (Moss and Lee, 2011, p. 331). Over nearly the same period of time (1997–2007), 91 elephants were speared, 57 of which died.

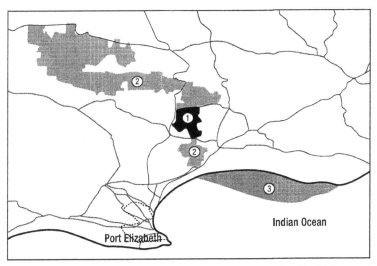

**FIGURE 14.4**  Expanded Addo Elephant National Park. (1) Original Addo elephant enclosure (approximately 220 km² or 80 mile²); (2) inland expansion (approximately 1200 km² or 430 mile²); and (3) coastal parkland and ocean including Stag and Bird Island (approximately 1400 km² or 500 mile²). (See also Figure 1.2 in Chapter 1.)

**TABLE 14.1** Annual numbers of human deaths caused by elephants in protected areas, communal lands, and enterprises using tame elephants (e.g., elephant theme parks and elephant-back safari operations) in South Africa

| Year | Protected Areas | Communal Land | Enterprise Using Tame Elephants | Total |
|---|---|---|---|---|
| 2002 | 2 | 1 | 0 | 3 |
| 2003 | 3 | 2 | 0 | 5 |
| 2004 | 2 | 0 | 0 | 2 |
| 2005 | 4 | 0 | 1 | 5 |
| 2006 | 2 | 0 | 1 | 3 |
| 2007 | 0 | 1 | 0 | 1 |
| Total number | 13 | 4 | 2 | 19 |
| Percent | 60% | 21% | 11% | |

*After Scholes and Mennell (2008, p. 222). Permission courtesy of the University of the Witwatersrand Press and the authors Scholes and Mennell.*

Despite the close contact and resulting conflict, when resources are shared elephants have not aggressively attacked humans. In fact, as shown earlier, they recognize and can discriminate between humans who pose a threat and those who do not. Their response is to avoid rather than seek conflict. In the Amboseli study, which through force of circumstances has had to deal with elephant–people conflict, solutions have been found and the prospects for mutual survival exist, albeit resting in the hands of humans.

The second possible hope that humans and elephants can coexist lies in the evolutionary history of humans and elephants. For a far greater part of this history or about 99.99% of the time, humans and elephants have coexisted with little or no conflict. In fact, humans probably feared elephants much more than the other way around. Only for a fraction of their mutual history, no more than one-hundredth of one percent of the time, were humans and elephants in mortal combat. It may be just possible that with the smallest sign of tolerance exhibited by humans for the well-being of elephants, it will be the elephants that meet us more than halfway.

Both solutions, however, may fail in the face of growing inhumanity and human greed. Unless a concerted international response is mounted within the immediate future, there is a real danger that the poaching of elephants and rhinoceros, for their tusks and horns, will eradicate both species from the wild.

Poaching must be viewed by the world's community in the identical light as we would a crime against humanity. We must see the slaughter of elephants, and other wildlife, with the same horror that we view ethnic cleansing. This slaughter of animals when seen in proportion to the global numbers involved far exceeds any assault humans have ever perpetrated upon their own kind.

To halt this Faustian tragedy, the international community, including those nations implicated in the crime, needs to take action simultaneously on two fronts: ban the sale of ivory and horn and vigorously and effectively suppress poaching.

Both actions have challenged global conservation forces in the past. Banning the sale of ivory has been effective; suppressing poaching much less so. International action can severely curtail the sale of ivory. Stopping poaching will require an advanced, well-equipped task force. Such a special force cannot be deployed simultaneously in all regions where poaching is rife. Instead, it should be deployed in a limited area where the chances of success are high, and, more importantly, the message to all poachers and those who support them will be the strongest. Once the suppression of poaching in one region has succeeded, methodology, personnel, and equipment can be transferred to cover a second selected region. Determined action of this kind would, in conjunction with the enforcement of trade agreements, rapidly reduce the slaughter to sustainable levels. If the account of elephants in this book is to have any meaning, then it calls upon our species to act with determination on behalf of our sorely tried brethren.

Having done so, however, the battle will not have been won. Rather, the real battle for mutual survival on this planet can then begin in earnest. This battle will be to confront the combined demands of a growing global human population and its inevitable drain upon the finite resources of the planet. Our inability to limit contamination and pollution of the fluid systems of the earth, air, and water is evident in almost every corner of the world. Careful studies in regions that, until very recently, were thought to be immune or too remote to suffer any serious degradation are now showing disturbing responses in vegetation and species degeneration. Hutto (2014) has shown that mule deer, mountain goats, and elk in remote regions of the high Rockies may all be in an irreversible decline.

Africa, despite environmental limitations discussed earlier, is seen by global and indigenous economists as the continent of the future. To hope that humans, in this diverse ecological continent, will be persuaded by the costly lessons we have learned to seek some equilibrium in which species other than ourselves might survive or even prosper is perhaps to hope in vain. All that has been recounted in this book, however, denies capitulation. At some point with this and many, many more efforts we, and elephants, as sentient beings, must prevail.

# References

Anthony, L., Spence, G., 2009. The Elephant Whisperer: My Life with the Herd in the African Wild. Thomas Dunne Books, St. Martin's Press, New York, 368 p.

Bates, L.A., et al., 2007. Elephants classify human ethnic groups by odor and garment color. Curr. Biol. 17, 1938–1942.

Bates, L.A., et al., 2008a. Do elephants show empathy? J. Concious. Stud. 15, 204–225.

Bates, L.A., et al., 2008b. African elephants have expectations of out-of-sight family members. Biol. Lett. 4, 34–36.

Beaune, D., et al., 2013. Seed dispersal strategies and the threat of defaunation in a Congo forest. Biodivers. Conserv. 22, 225–238.

Bekoff, M., 2002. Minding Animals: Awareness, Emotions and Heart. Oxford University Press, Oxford.

Bekoff, M., Pierce, J., 2009. Wild Justice: The Moral Lives of Animals. University of Chicago Press, Chicago, 204 p.

Bell, W.D.M., 1960. Bell of Africa. Neville Spearman and The Holland Press, London, 228 p.

Bradshaw, G.A., 2009. Elephants on the Edge: What Animals Teach Us About Humanity. Sheridan Books, Ann Arbor, MI, 310 p.

Brenner, E., et al., 2006. Plant neurobiology: an integrated view of plant signaling. Trends Plant Sci. 11, 413–419.

Brown, C.H., 1994. Sound localization. In: Fay, R.R., Popper, A.N. (Eds.), Comparative Hearing: Mammals. Springer, Berlin, Heidelberg, New York, pp. 57–96.

Buckley, C., 2009. Tarra & Bella. G.P. Putnam's Sons, Penguin Group, Inc., New York.

Byrne, R., 1997. What's the use of anecdotes? Distinguishing psychological mechanisms in primate tactical deception. In: Mitchell, R.W., Thompson, N.S., Miles, H.L. (Eds.), Anthropomorphism, Anecdotes, and Animals. State University of New York Press, New York, pp. 134–150 (chapter 12).

Byrne, R., 2006. Animal cognition: know your enemy. Curr. Biol. 16, R686–R688.

Byrne, R., Bates, L., 2006. Why are animals cognitive? Curr. Biol. 16, R445–R448.

Byrne, R., Bates, L., 2009. Elephant cognition in primate perspective. Comp. Cogn. Behav. Rev. 4, 65–79.

Byrne, R., Bates, L., 2010. Primate social cognition: uniquely primate, uniquely social or just unique. Neuron 65, 815–830.

Byrne, R., Bates, L., 2011a. Elephant cognition: what we know about what elephants know. In: Moss, C., Croze, H., Lee, P. (Eds.), The Amboseli Elephants: A Long-Term Perspective of a Long-Lived Animal. University of Chicago Press, Chicago, pp. 174–181 (chapter 11).

Byrne, R., Bates, L., 2011b. Cognition in the wild: exploring animal minds with observational evidence. Biol. Lett. 7, 619–622.

Byrne, R.W., et al., 2009. How did they get here from there? Detecting changes in direction in terrestrial ranging. Anim. Behav. 77, 619–631.

Campos-Arceiz, A., Blake, S., 2011. Megagardeners of the forest—the role of elephants in seed dispersal. Acta Oecol. 37, 542–553.

Caro, T.M., Hauser, M.D., 1992. Is there teaching in non-human animals? Q. Rev. Biol. 67, 151–174.

Carter, P., 2008. The elephants of Mfuwe Lodge. http://www.africatravelguide.com/articles/the-elephants-of-mfuwe-lodge.html.

Chase, M., 2007. Elephants learn to avoid land mines in war torn Angola. Reported by L. Marshall, National Geographic News. NationalGeographic.comwww.elephantswithoutborders.org.

Chase, M., Griffin, C., 2011. Elephants of southeast Angola in war and peace: their decline, re-colonization and recent status. Afr. J. Ecol. 49, 353–361.

Chiyo, P.I., et al., 2011. Using molecular and observational techniques to estimate the number and raiding patterns of crop-raiding elephants. J. Appl. Ecol. 48, 788–796.

Christo, C., Wilkinson, W., 2009. Walking Thunder: In the Footsteps of the African Elephant. Merrill Publishers Ltd, London, New York, 158 p.

Clemins, P.J., et al., 2005. Automatic classification and speaker identification of African elephant (*Loxodonta africana*) vocalizations. J. Acoust. Soc. Am. 117, 1–8.

Cochrane, E.P., 2003. The need to be eaten: Balanites wilsoniana with and without elephant seed-dispersal. J. Trop. Ecol. 19, 579–589.

Darwin, C., 1897. The Expression of the Emotions in Man and Animals. D. Appleton and Co., New York, 372 p.

Dawkins, R., 1989. The Selfish Gene. Oxford University Press, New York (paperback edition, first publication 1976), 352 p.

Dawkins, R., 2000. Unweaving the Rainbow. First Mariner Books (original publication 1998), 337 p.

Dawkins, R., 2008. The Extended Phenotype. Oxford University Press, New York, (Original publication 1982), 313 p.

Dawkins, R., Krebs, J.R., 1978. Animal signals: information or manipulation? In: Krebs, J.R., Davies, N.B. (Eds.), Behavioral Ecology: An Evolutionary Approach. Blackwell Scientific Publications, Oxford, pp. 282–309.

de Waal, F.B.M., 1996. Good Natured: The Origins of Right and Wrong in Humans and Other Animals. Harvard University Press, Cambridge.

de Waal, F.B.M., 2008. Putting altruism back into altruism: the evolution of empathy. Annu. Rev. Psychol. 59, 279–300.

de Waal, F.B.M., 2013. The Bonobo and the Atheist: In Search of Humanism Among the Primates. W.W. Norton and Co., New York, 289 p.

Dieudonne, J., 1998. Keystone species as indicators of climate change. MS Thesis, University of Virginia, 147 p.

Dixon, B.A., 2008. Animals, Emotion and Morality. Prometheus Books, New York, 281 p.

Douglas-Hamilton, I., et al., 2006. Behavioral reactions of elephants towards a dying and deceased matriarch. Appl. Anim. Behav. Sci. 100, 87–102.

Eagleman, D.M., 2011. Incognito: The Secret Lives of the Brain. Pantheon Books, New York, 289 p.

Evans, K.E., Harris, S., 2008. Adolescence in male African elephants, *Loxodonta africana*, and the importance of sociality. Anim. Behav. 76, 779–787.

Fitch, W.T., 1997. Vocal tract length and formant frequency dispersion correlate with body size in rhesus macaques. J. Acoust. Soc. Am. 102, 1213–1222.

Fitch, W.T., 2000. Skull dimensions in relation to body size in nonhuman mammals: the causal bases for acoustic allometry. Zoology 103, 40–58.

Fitch, W.T., Hauser, M.D., 2002. Unpacking "honesty" vertebrate vocal production and the evolution of acoustic signals. In: Acoustic Communication, Simmons, A., Fay, R.R., Popper, N.

(Eds.), Springer Handbook of Auditory Research, vol. 16. Springer-Verlag, Berlin, Heidelberg, New York, p. 424.

Fitch, W.T., Reby, D., 2001. The descended larynx is not uniquely human. P. Roy. Soc. Lond. B Bio. 268, 1669–1675.

Foer, J., 2011. Moonwalking with Einstein: The Art and Science of Remembering Everything. Penguin Press, New York, 307 p.

Foley, C., et al., 2008. Severe drought and calf survival in elephants. Biol. Lett. 4, 541–544.

Franke, S.J., Swanson Jr., G.W., 1989. A brief tutorial on the fast field program (FFP) as applied to sound propagation in the air. Appl. Acoust. 27, 203–215.

Garstang, M., 2004. Long-distance, low-frequency elephant communication. J. Comp. Physiol. A 190, 791–805.

Garstang, M., 2009. Precursor tsunami signals detected by elephants. The Open Conserve. Biol. J. 3, 1–3.

Garstang, M., Fitzjarrald, D., 1999. Observations of Surface to Atmosphere Interactions in the Tropics. Oxford University Press, New York, 405 p.

Garstang, M., et al., 1995. Atmospheric controls on elephant communication. J. Exp. Biol. 198, 939–951.

Garstang, M., et al., 2005. The daily cycle of low-frequency elephant calls and near-surface atmospheric conditions. Earth Interact. 9, 1–21.

Garstang, M., et al., 2014. Response of African elephants (*Loxodonta africana*) to seasonal changes in rainfall. PLoS One 9, e108736, 13 p.

Gould, J.L., 2004. Animal cognition. Curr. Biol. 14, 372–375.

Gröning, K., Saller, M., 1999. Elephants: A Cultural and Natural History. Könemann Verlagsgesell-shaft mbH, Cologne, 482 p. (English edition, original publication 1998).

Gwynne, S.G., 2010. Empire of the Summer Moon. Scribner, a Division of Simon and Schuster, Inc., New York, p. 260.

Hägstrum, J.T., 2000. Infrasound and the avian navigational map. J. Exp. Biol. 203, 1103–1111.

Hamlin, J.K., Wynn, K., Bloom, P., 2007. Social evaluation by pre-verbal infants. Nature 450, 557–560.

Heffner, R., Heffner, H., 1982. Hearing in the elephant (*Elephas maximus*): absolute sensitivity, frequency discrimination, and sound localization. J. Comp. Physiol. 96, 926–944.

Heffner, R., Heffner, H., 1984. Sound localization in large mammals: localization of complex sounds by horses. Behav. Neurosci. 98, 541–555.

Herbst, C.T., et al., 2013. Complex vibratory patterns in an elephant larynx. J. Exp. Biol. 216, 4054–4064.

Heyes, C.M., 1993. Anecdotes, training, trapping and triangulating: do animals attribute mental states? Anim. Behav. 46, 177–188.

Holdrege, C., 2001. Elephantine intelligence. Nature Institute Monographs, Context #5 The Nature Institute, New York, 6 p.

Hutto, J., 2006. Illumination in the Flatwoods: A Season Living Among the Wild Turkey, second ed. Lyons Press, Guilford, CT, paperback, 240 p.

Hutto, J., 2014. Touching the Wild: Living with the Mule Deer of Deadman Gulch. Skyhorse Publishing, New York, 306 p.

Johanson, D., Edgar, B., 1996. From Lucy to Language. Wits University Press, Johannesburg, 272 p.

Kelley, M., Garstang, M., 2013. On the possible detection of lightning storms by elephants. Animals 3, 349–355.

Krauss, R.M., 1987. The role of the listener: addressee influences on message formulation. J. Lang. Soc. Psychol. 6, 81–97.

Kuhn, G.F., 1977. Model for the interaural time differences in the azimuthal plane. J. Acoust. Soc. Am. 62, 157–167.

Kuhn, G.F., 1987. Physical acoustic and measurements pertaining to directional hearing. In: Yost, W.A., Gourevitch, G. (Eds.), Directional Hearing. Academic Press, New York, pp. 305.

Langbauer Jr., W.R., et al., 1991. African elephants respond to distant playback of low-frequency conspecific calls. J. Exp. Biol. 157, 35–46.

Larom, D., et al., 1997. The influence of surface atmospheric conditions on the range and area reached by animal vocalizations. J. Exp. Biol. 200, 421–431.

Lee, P.C., Moss, C.J., 2012. Wild female African elephants (*Loxodonta africana*) exhibit personality traits of leadership and social integration. J. Comp. Physiol. 126, 224–232.

Lee, P.C., Poole, J., 2011. Reproductive strategies and social relationships. Part 4. In: Moss, C.J., Croze, H., Lee, P.L. (Eds.), The Amboseli Elephants: A Long-Term Perspective on a Long-Lived Mammal. Chicago Press, Chicago, IL, pp. 187–291.

Lee, S.W., et al., 1986. Impedance formulation of the fast field program for acoustic wave propagation in the atmosphere. J. Acoust. Soc. Am. 79, 628–634.

Leggett, K.E.A., et al., 2011. Matriarchal associations and reproduction in a remnant sub-population of desert dwelling elephants in Namibia. Pachyderm 49, 20–32.

Leong, K.M., Ortolani, A., Burks, K.D., Mellen, J.D., Savage, A., 2003. Quantifying acoustic and temporal characteristics of vocalizations for a group of captive African elephants (*Loxodonta africana*). Bioacoustics 13, 213–231.

LePort, A.K., et al., 2012. Behavioral and neuroanatomical investigation of highly superior autobiographical memory (HSAM). Neurobiol. Learn. Mem. 96, 78–92.

Levenson, R.W., 2003. Blood, sweat and fears: the autonomic architecture of emotion. Ann. N.Y. Acad. Sci. 1000, 348–366.

Lindeque, M., Lindeque, P.M., 1991. Satellite tracking of elephants in northwestern Namibia. Afr. J. Ecol. 29, 196–206.

Lister, A.M., 2013. The role of behavior in adaptive morphological evolution of African proboscideans. Nature 500, 331–334.

London Daily Telegraph, 2012. Elephant whisperer saved Baghdad zoo. London Daily Telegraph (March 13). http://www.telegraph.co.uk/news/obituaries/9131585/Lawrence-Anthony.html.

Long, G.R., 1994. Psychoacoustics. In: Fay, R.R., Popper, A.N. (Eds.), Comparative Hearing: Mammals. Springer, Berlin, Heidelberg, New York, pp. 18–56.

Masson, J.M., McCarthy, S., 1995. When Elephants Weep: The Emotional Lives of Animals. Delacorte Press Bantam Doubleday Dell Pub. Group Inc., New York, 291 p.

McCarthy, T., Rubidge, B., 2005. The Story of Earth & Life: A Southern African Perspective on a 4.6-Billion-Year Journey. Random House Struik, Cape Town, 335 p.

McComb, K.E., 1991. Female choice for high roaring rate in red deer, *Cervus elaphus*. Anim. Behav. 41, 79–88.

McComb, K., et al., 2000. Unusually extensive networks of vocal recognition in African elephants. Anim. Behav. 59, 1103–1109.

McComb, K., et al., 2001. Matriarchs as repositories of social knowledge in African elephants. Science 292, 491–494.

McComb, K., et al., 2011. Leadership in elephants: the adaptive value of age. P. Roy. Soc. B Bio. 278, 3270–3276.

McComb, K., et al., 2003. Long distance communication of acoustic cues to social identity in African elephants. Anim. Behav. 65, 317–329.

McGaugh, J.L., 2003. Memory and Emotion: The Making of Lasting Memories. Columbia University Press, New York, 162 p.

Mendoza, S.P., Ruys, J.D., 2001. The beginning of an alternative view of the neurobiology of emotion. Soc. Sci. Inform. 40, 39–60.

Moss, C.J., 1988. Elephant Memories: Thirteen Years in the Life of an Elephant Family. Ballentine Books, New York, 335 p.

Moss, C., Lee, P., 2011. Female social dynamics: fidelity and flexibility. In: Moss, C.J., Croze, H., Lee, P. (Eds.), The Amboseli Elephants: A Long-Term Perspective on a Long-Lived Animal. University of Chicago Press, Chicago, pp. 205–208 (chapter 13).

Mutwa, V.C., 1965. Indaba My Children. Blue Crane Books, Johannesburg, 562 p.

Myhrvold, C.L., et al., 2012. What is the use of elephant hair. PLoS One 7, e47018.

Nicol, C., 2013. Do elephants have souls? The New Atlantis. J. Tech. Soc. 38, 10–70.

O'Connell-Rodwell, C.E., et al., 2000. Seismic properties of Asian elephant (*Elephas maximum*) vocalizations and locomotion. J. Acoust. Soc. Am. 108, 3066–3072.

O'Connell-Rodwell, C.E., et al., 2001. Exploring the potential use of seismic waves as a communication channel by elephants and other large mammals. Am. Zool. 41, 1157–1170.

O'Connell-Rodwell, C.E., et al., 2004. Interactive patterns of vocal communication in African elephant herds. J. Acoust. Soc. Am. 115, 2555.

O'Connell-Rodwell, C.E., et al., 2007. Wild African elephants (*Loxodonta africana*) discriminate between familiar and unfamiliar conspecific seismic alarm calls. J. Acoust. Soc. Am. 122, 823–830.

O'Connell-Rodwell, C.E., et al., 2012. Antiphonal vocal bouts associated with departures in free-ranging African elephant family groups (*Loxodonta africana*). Bioacoustics 21, 215–224.

Owings, D.H., Morton, E.S., 1997. The role of information in communication on assessment/management approach. In: Owings, D.H., Beecher, M.D., Thompson, N.S. (Eds.), Perspectives in Ethology, Vol. 12: Communication. Plenum, New York, pp. 359–390.

Owings, D.H., Morton, E.S., 1998. Animal Vocal Communication: A New Approach. Cambridge University Press, Cambridge, 284 p.

Palmer, A.R., 2004. Reassessing mechanisms of low-frequency sound localization. Curr. Opin. Neurobiol. 14, 457–460.

Parker, E.S., et al., 2006. A case of unusual autobiographical remembering. Neurocase 12, 35–49.

Patterson, G., 2009. The Secret Elephants: The Rediscovery of the World's Southerly Elephants. Penguin Books, Johannesburg, 218 p.

Payne, K., 1998. Silent Thunder: In the Presence of Elephants. Simon and Schuster, New York, p. 288.

Peterson, D., 2011. The Moral Lives of Animals. Bloomsbury Press, New York, 342 p.

Pierce, A.D., 1981. Acoustics: An Introduction to Its Physical Principles and Applications. McGraw-Hill, New York, p. 61.

Pijanowski, B.C., et al., 2011. Soundscape ecology: the science of sound in the landscape. Bioscience 61, 203–216.

Plotnik, J.M., et al., 2011. Elephants know when they need a helping trunk in a cooperative task. Proc. Natl. Acad. Sci. U. S. A. 108, 5116–5121.

Poole, J.H., 1996. Coming of Age with Elephants. Hyperion, New York, 288 p.

Poole, J.H., 2011. Behavioral contexts of elephant acoustic communication. In: Moss, C.J., Croze, H., Lee, P. (Eds.), The Amboseli Elephants: A Long-Term Perspective on a Long-Lived Mammal. University of Chicago Press, Chicago, IL, pp. 125–161 (chapter 9).

Poole, J.H., Granli, P., 2004. The visual, tactile and acoustic signals of play in African savanna elephants. Elephant Voices & Amboseli Research Project Report, 1–7.

Poole, J.H., et al., 1988. The social context of some very low-frequency calls of African elephants. Behav. Ecol. Sociobiol. 22, 385–392.

Poole, J., et al., 2006. Vocal imitation in African savannah elephants (*Loxodonta africana*). Razprave IV Razreda Sazu X LVII-3, 117–124.

Raspet, R., et al., 1985. A fast field program for sound propagation in a layered atmosphere above an impedance ground. J. Acoust. Soc. Am. 77, 345–352.

Reby, D., McComb, K., 2003. Anatomical constraints generate honesty: acoustic cues to age and weight in roars of red deer stags. Anim. Behav. 65, 519–530.

Regan, T., 2004. The Case for Animal Rights. University of California Press, Berkeley, Los Angeles, 425 p.

Rendall, D., 1996. Social communication and vocal recognition in free-ranging rhesus monkeys (*Macaca mulatta*). PhD dissertation, University of California.

Riede, T., Fitch, W.T., 1999. Vocal tract length and acoustics of vocalization in the domestic dog, *Canis familiaris*. J. Exp. Biol. 202, 2859–2869.

Roderigues, J., 1992. The Game Rangers. J.A. Roderigues, Innesdale, South Africa, 215 p.

Rosowski, J.J., 1994. Outer and middle ears. In: Fay, R.R., Popper, A.N. (Eds.), Comparative Hearing: Mammals. Springer, Berlin, Heidelberg, New York, pp. 172–247.

Rowlands, M., 2012. Can Animals Be Moral. Oxford University Press, New York, 259 p.

Scholes, R.J., Mennell, K.G. (Eds.), 2008. Elephant Management: A Scientific Assessment for South Africa. Wits University Press, Johannesburg, 586 p.

Seyfarth, R.M., Cheney, D.L., 2003a. Meaning and emotion in animal vocalization. Ann. N.Y. Acad. Sci. 1000, 32–55.

Seyfarth, R.M., Cheney, D.L., 2003b. Signalers and receivers in animal communication. Annu. Rev. Psychol. 54, 145–173.

Seyfarth, R.M., et al., 2010. The central importance of information in studies of animal communication. Anim. Behav. 80, 3–8.

Shannon, G., et al., 2013. Effects of social disruption in elephants persist decades after culling. Front. Zool. 10, 62.

Sheldrick, D., 2012. Love, Life and Elephants: An African Story. Farrar Straus and Giroux Publishers, New York, 334 p.

Shoshani, J., 2002. Proboscidea (elephants). In: Encyclopedia of Life Sciences. Nature Publishing Group.

Shoshani, J., et al., 2006. Elephant brain. Part I: gross morphology, functions, comparative anatomy and evolution. Brain Res. Bull. 70, 124–127.

Smet, A.F., Byrne, R.W., 2013. African elephants can use human pointing cues to find hidden food. Curr. Biol. 23, 2033–2037.

Soltis, J., 2010. Vocal communication in African elephants (*Loxodonta africana*). Zoo Biol. 29, 192–209.

Soltis, J., 2013. Emotional communication in African elephants. In: Altenmüller, E., Schmidt, S., Zimmerman, E. (Eds.), The Evolution of Emotional Communication. Oxford University Press, New York, NY, pp. 106–115 (chapter 7).

Southworth, M., 1969. The sonic environment of cities. Environ. Behav. 1, 49–70.

Stander, P.E., Stander, J., 1988. Characteristics of lion roars in Etosha National Park. Madoqua 13, 315–318.

Stevenson-Hamilton, J., 1993. South African Eden. Struik Publishers, Cape Town, 336 p.

Stoeger, A.S., et al., 2012. An Asian elephant imitates human speech. Curr. Biol. 22, 2144–2148.

Thornton, A., Raihani, J.J., 2008. The evolution of teaching. Anim. Behav. 75, 1823–1836.

Titze, I., 1994. Principles of Voice Production. Prentice-Hall, Englewood Cliffs.

Tyson, P.D., 1986. Climatic Changes and Variability in Southern Africa. Oxford University Press, Cape Town.

van Garde, R.J., 2009. Elephants: Facts and Fables. International Fund for Animal Welfare RPT, Cape Town, South Africa. www.ifaw.org.

von Békésy, G., 1960. Experiments in Hearing. McGraw-Hill, New York, p. 745 (translated and edited by E.G. Werer).

Watson, L., 2003. Elephantoms: Tracking the Elephant. Penguin Books, South Africa, Struik Publishers, 256 p. (first published 2002, W.W. Norton and Co., New York).

Weissenböck, N.M., et al., 2012. Taking the heat: thermoregulation in Asian elephants under different climatic conditions. J. Comp. Physiol. B 182, 311–319.

Weissenböck, N.M., et al., 2010. Thermal windows on the body surface of African elephants (*Loxodonta africana*) studied by infrared thermography. J. Therm. Biol. 35, 182–188.

Willmer, P., et al., 2000. Environmental Physiology of Animals. Blackwell, Oxford, p. 644.

Wilson, M.L., et al., 2001. Does participation in intergroup conflict depend on numerical assessment, range location or rank for wild chimpanzees? Anim. Behav. 61, 1203–1216.

Wittemyer, G., et al., 2005. The socioecology of elephants: analysis of the processes creating multitiered social structures. Anim. Behav. 69, 1357–1371.

Wittemyer, G., et al., 2009. Where sociality and relatedness diverge: the genetic basis for hierarchical social organization in African elephants. Proc. Biol. Sci. 276, 3513–3521.

Wynn, K., 2008. Some innate foundations of social and moral cognition. In: Carruthers, P., Laurence, S., Stich, S. (Eds.), The Innate Mind: Foundations and Future. Oxford University Press, Oxford, pp. 330–347.

# Supplemental Bibliography

Research in elephant behavior is progressing rapidly. Papers and reports published since completing the text and of interest to the work are listed below but not cited in the text.

## COMMUNICATION

Partan, S.R., 2013. Ten unanswered questions in multimodal communication. Behav. Ecol. Sociobiol. 67, 1523–1539.

Stoeger, A.S., de Silva, S., 2014. African and Asian elephant vocal communication: a cross-species comparison. In: Witzany, G. (Ed.), Biocommunication of Animals. Springer Science, Dordrecht, pp. 21–39 (chapter 3).

Stoeger, A.S., Zeppelzauer, M., Baotic, A., 2014. Age-group estimation in free-ranging African elephants based on acoustic cues of low-frequency rumbles. Bioacoustics 23, 231–246.

Zeppelzauer, M., Hensman, S., Stoeger, A.S., 2014. Towards an automated acoustic detection system for free-ranging elephants. Bioacoustics 23. http://dx.doi.org/10.1080/09524622.2014.906321.

## BEHAVIOR

de Silva, S., 2014. How does empathy help elephants? http://asianelephant.wordpress.com/2014/03/03/how-does-empathy-help-elephants/.

King, L.E., 2013. Elephants and bees. Sanctuary Asia (October). http://www.sanctuaryasia.com/ConflictResolution.

King, L.E., Douglas-Hamilton, I., Vollrath, F., 2011. Beehive fences as effective deterrents for crop-raiding elephants: field trials in northern Kenya. Afr. J. Ecol. 49, 431–439.

Kuiper, T.R., Parker, D.M., 2014. Elephants in Africa: big, grey biodiversity thieves. S. Afr. J. Sci. 110, 7–9.

Landman, M., Kerley, G.I., 2014. Elephants both increase and decrease availability of browse resources for black rhinoceros. Biotropica 46, 42–49.

Morell, V., 2014. It's time to accept that elephants, like us, are empathetic beings. http://news.nationalgeogrpahic.com/news/2014/02/140221-elephants-poaching-empathy-grief/.

Plotnik, J.M., de Waal, F.B.M., 2014. Asian elephants (*Elephas maximum*) reassure others in distress. Peer J 2, e278.

Shannon, G., Slotow, R., Durant, S.M., Sayialel, K.N., Poole, J., Moss, C., McComb, K., 2013. Effects of social disruption in elephants persist decades after culling. Front. Zool. 10, 62.

Soltis, J., King, L.E., Douglas-Hamilton, I., Vollrath, F., Savage, A., 2014. African elephant alarm calls distinguish between threats from humans and bees. PLoS One 9, e89403.

## INTELLIGENCE

Borenstein, S., 2014. Elephants prove discerning listeners of us humans. http://bigstory.ap.org/article/elephants-prove-discerning-listeners-us-humans.

Jabr, F., 2014. Searching for the elephant's genius inside the largest brain on land. Sci. Am. (February). http://blogs.scientificamerican.com/brainwaves/2014/02/26/searching-for-the-elephants-genius-inside-the-largest-brain-on-land/.

## MOVEMENT

Barua, M., 2013. Circulating elephants: unpacking the geographies of a cosmopolitan animal. Trans. Inst. British Geograph. 39, 559–573.

Jachowski, D.S., Montgomery, R.A., Slotow, R., Millspaugh, J.J., 2013. Unravelling complex associations between physiological state and movement of African elephants. Funct. Ecol. 27, 1166–1175.

## LEARNING

Stoeger, A.S., Manger, P., 2014. Vocal learning in elephants: neural bases and adaptive context. Curr. Opin. Neurobiol. 28, 101–107.

# Index

Note: Page numbers followed by *f* indicate figures and *t* indicate tables.

Printed in the United States
By Bookmasters